Sources and Studies in the History of Mathematics and Physical Sciences

Sources and Studies in the History of Mathematics and Physical Sciences was inaugurated as two series in 1975 with the publication in Studies of Otto Neugebauer's seminal three-volume History of Ancient Mathematical Astronomy, which remains the central history of the subject. This publication was followed the next year in Sources by Gerald Toomer's transcription, translation (from the Arabic), and commentary of Diocles on Burning Mirrors. The two series were eventually amalgamated under a single editorial board led originally by Martin Klein (d. 2009) and Gerald Toomer, respectively two of the foremost historians of modern and ancient physical science. The goal of the joint series, as of its two predecessors, is to publish probing histories and thorough editions of technical developments in mathematics and physics, broadly construed. Its scope covers all relevant work from pre-classical antiquity through the last century, ranging from Babylonian mathematics to the scientific correspondence of H. A. Lorentz. Books in this series will interest scholars in the history of mathematics and physics, mathematicians, physicists, engineers, and anyone who seeks to understand the historical underpinnings of the modern physical sciences. All works are reviewed to meet the highest standards of scientific literature.

Titles from this series are indexed by Scopus.

Clemency Montelle • Kim Plofker • Glen Van Brummelen

From Samarqand to Jaipur

Evolving al-Kāshī's Approximation to the Sine of One Degree

Springer

Clemency Montelle
School of Mathematics and Statistics
University of Canterbury
Christchurch, New Zealand

Kim Plofker
Department of Mathematics
Union College
Schenectady, NY, USA

Glen Van Brummelen
Department of Mathematical Sciences
Trinity Western University
Langley, BC, Canada

ISSN 2196-8810 ISSN 2196-8829 (electronic)
Sources and Studies in the History of Mathematics and Physical Sciences
ISBN 978-3-031-89086-4 ISBN 978-3-031-89087-1 (eBook)
https://doi.org/10.1007/978-3-031-89087-1

This Springer imprint is published by the registered company Springer Nature Switzerland AG
The registered company address is: Gewerbestrasse 11, 6330 Cham, Switzerland

To David Pingree, teacher and guide, whose pioneering research into the cross-cultural circulation of the exact sciences throughout history continues to inspire.

Preface

This project commenced when the three of us met together at the Chennai Mathematical Institute in south India for a month-long intensive research visit in 2014. Little did we anticipate that what started off as a brief examination of a curious manuscript with fascinating diagrams and distinctive symbolism would lead to such a deep and expansive scholarly investigation into one of the most gripping episodes in the history of trigonometry, spanning at least four different cultures of inquiry over a millennium and a half. We are grateful for the CMI's generous hospitality hosting us as visiting scholars, as well as the Rutherford Discovery Fellowship funding from the Royal Society of New Zealand that supported this scholarly exchange.

The month was well spent. It quickly became clear that the manuscript contained a record of investigations into the challenging problem of finding an accurate value for the sine of one degree, the fundamental constant underlying all trigonometric tables, and hence all of mathematical astronomy. The narrative in this manuscript begins with reports of early 15th-century scholar Jamshīd al-Kāshī's innovative method, converting the problem to a cubic equation and using an iterative method to solve it. A description of the work of his Samarqand patron Ulugh Beg follows. To find this work in Sanskrit in 18th-century Jaipur is already remarkable. However, this treatise also goes on to describe further developments by two astronomers in Jai Singh's court, showing that this mathematical tradition was being actively engaged and extended three centuries after al-Kāshī, in India.

The project continued and expanded over the next few years as new sources came to light and the line of inquiry expanded. We are indebted to the University of Canterbury Erskine Fellowship program that supported further visits in subsequent years to continue the research and finish the project.

We are also deeply grateful to the staff at the Shri Ram Charan Sharma Museum of Indology, Jaipur and the Tehran Majlis digital library, who kindly provided us with copies of the manuscripts included in this study. A special thanks in particular goes to Professor Peyman Zawar-Reza, who assisted us with procuring copies of the Persian manuscripts.

Many other people have supported the research and helped this book come to fruition. First, we thank Rev Dr Jambugahapitiye Dhammaloka and Sahana Cidambi who joined us for a reading group with many questions and insights. We are also grateful to Jan Hogendijk who read an early draft of the work and provided numerous valuable comments, and likewise to Muhammad Ziayan Omer for his assistance in editing and computation. We thank Maayan Levy for her editorial support, giving us invaluable assistance in the final stages of preparing the manuscript for publication.

We also express our hearty thanks to the team at Springer Nature, especially Associate Editor Nicholas DiBenedetto and Production Editor Vignesh Vishwanathan, for their help and support in navigating the process from contract to production.

Finally, we thank our family and friends for their continuing support of us in the surprising directions and fascinating places our curiosity and research took us.

Contents

Chapter 1
Introduction: The Sine of 1° and its History

In the first section of this chapter (1.1) we sketch the history of techniques for computing the sine of 1° in Hellenistic trigonometry and its Arabic, Persian and European successors, along with the mathematical foundations of iterative techniques for this calculation. Section 1.2 situates this problem in the context of Indian trigonometry, and section 1.3 discusses in more detail the Sanskrit text and manuscript that form the main subject of this study.

In chapter 2 we edit and translate the two recensions of the manuscript's surviving prose text. We analyze its mathematical content in chapter 3 and reproduce and check all the numerical computations. We then step back from this text to trace (in part) the textual history of the mathematical methods contained therein: first in Islamic mathematical traditions (chapter 4 and Appendix A) and then in Sanskrit ones (chapter 5). Finally, chapter 6 summarizes our findings and their implications for the history of early modern Islamic and Indian mathematical sciences.

1.1 The Sine of 1° in Hellenistic, Islamic and European Mathematics

It has been known since antiquity that the trigonometric chords or sines[1] of certain arcs or angles can be determined with geometric precision, while others can be computed only approximately. In particular, the chord or sine of any integer multiple of 3°, or of any angle derived from any such multiple by repeated bisection, can be found exactly. But it is theoretically impossible using Euclidean tools to trisect an arbitrary angle. Hence, knowing the chord or sine of 3° is useless for the purpose of determining the chord or sine of 1° if one stays within the bounds of classical geometry. This proved to be a serious obstacle in trigonometric computation, since a value for the chord or sine of 1° is foundational for constructing tables of trigonometric values for each degree of arc.

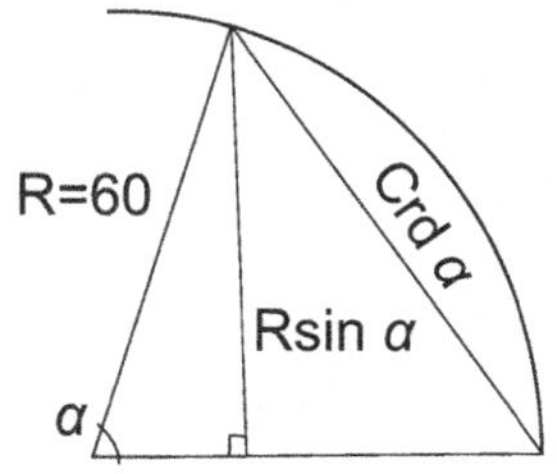

Figure 1.1: The chord (Crd) and the scaled sine or R sin of angle/arc α in a circle with trigonometric radius $R = 60$. The R sin of α is half the chord of 2α.

[1] Since the sine of any angle is half the chord of double that angle (see figure 1.1), and trigonometric identities permit the derivation of the sine of any angle from the known sine of its half or its double, the computation of chords differs only trivially from that of sines. Throughout this work, we denote the chord of an angle α in a circle of radius R by Crd α, and its sine scaled to the trigonometric radius R by $R \sin \alpha$.

C. Montelle et al., *From Samarqand to Jaipur*, Sources and Studies in the History of Mathematics and Physical Sciences, https://doi.org/10.1007/978-3-031-89087-1_1

1.1.1 Ptolemy's Approximation for the Chord of 1°

Hellenistic Greek trigonometry worked only with chords of arcs or angles in a circle of 360° and a standard radius of $R = 60$. Indian mathematician-astronomers around the start of the Common Era appear to have originated the idea of modifying these chords to the more convenient half-chords or sines (Sanskrit *jyā-ardha*, literally "half-bowstring", abbreviated to *jyā*, *jīvā*, and other synonyms for "bowstring"). They supplemented these sines by auxiliary functions such as the versed sine or versine (*utkrama-jyā*, "reversed sine"). Both the Hellenistic system of chords and the Indian system of sines were incorporated into medieval Islamic trigonometry and thence into its Latin counterpart, where the Arabized form *jayb* of Sanskrit *jīvā* was translated into its modern name *sinus* or "sine" (Van Brummelen 2009, pp. 137–138). But Islamic trigonometry remained strongly committed to the Hellenistic geometric methodology of Euclid and his successors, including near-universal adherence to the standard radius $R = 60$, which it adapted to all the trigonometric quantities then in use: sine, versine, cosine, and (eventually) tangent.

All ancient and medieval users of trigonometry employed some form of trigonometric tables, with function values tabulated at intervals of argument small enough to enable the calculation of acceptably accurate intermediate function values by interpolating between tabulated ones. For the Alexandrian astronomer Claudius Ptolemy in the second century CE, the desired table of chords for each successive degree of arc required determining the value of the chord of a single degree. He described in his Greek treatise *Mathematical Syntaxis* (called in Arabic *al-majisṭī* from its later Greek epithet *megistē* or "greatest", whence the modern name *Almagest*) a method of computing an approximate value of the chord of 1° by a double inequality (Toomer 1984, pp. 55–56). Briefly, Ptolemy's approximation relies on the following relation between arbitrary angles α and β in degrees:

$$0^\circ < \beta < \alpha < 180^\circ \quad \Longrightarrow \quad \frac{\operatorname{Crd}\alpha}{\operatorname{Crd}\beta} < \frac{\alpha}{\beta}. \tag{1.1}$$

From this inequality he deduced that

$$\frac{\operatorname{Crd}\frac{3}{2}^\circ}{\operatorname{Crd}1^\circ} < \frac{\frac{3}{2}^\circ}{1^\circ} \quad \text{and} \quad \frac{\operatorname{Crd}1^\circ}{\operatorname{Crd}\frac{3}{4}^\circ} < \frac{1^\circ}{\frac{3}{4}^\circ}, \quad \text{or} \quad \frac{2}{3}\operatorname{Crd}\frac{3}{2}^\circ < \operatorname{Crd}1^\circ < \frac{4}{3}\operatorname{Crd}\frac{3}{4}^\circ. \tag{1.2}$$

Using sexagesimal computations[2] with radius 60, Ptolemy found that $\operatorname{Crd}\frac{3}{2}^\circ \approx 1;34,15$ and $\operatorname{Crd}\frac{3}{4}^\circ \approx 0;47,8$. With creative rounding, he worked out both the upper and lower bounds in his double inequality to be 1;2,50, which he adopted as the value of Crd 1° itself.

This use of approximative methods within geometry was poorly received by some of Ptolemy's successors. For instance, in the early 11th century the Cairo scientist Ḥasan ibn al-Haytham remarked that Ptolemy's assertion that Crd 1° is simultaneously less than, greater than, and equal to 1;2,50 "is a contradictory statement. Besides, it is ugly and jars the ears, and no one is able to hear it" (Voss 1985, p. 28). In the 12th century this prompted the mathematician/physician Ibn Yaḥyā Samaw'al al-Maghribī to redefine the number of degrees in a circle to 480, thereby bypassing the problem altogether, since the sine of one unit in this system is equal to $\frac{1}{2}\operatorname{Crd}\frac{3}{2}^\circ$ (Van Brummelen, Mimura, and Kerai 2013).

Such objections did not deter several medieval Islamic and European scholars from further experimentation with Ptolemy's technique in order to improve his estimate. Since the chord or sine may be determined exactly for any angle that can be written in degrees as $3m/2^n$ for positive integers m and n, one may choose to work with such angles closer to 1° to obtain better bounds and a more accurate value. For instance, rather than the fractional degree amounts $\frac{3}{2}$ and $\frac{3}{4}$, the 10th-century Egyptian astronomer Abū al-Ḥasan ibn Yūnus used $\frac{9}{8}$ and $\frac{15}{16}$; in the 14th century, Richard of Wallingford discussed extending the procedure to $\frac{1026}{1024}$ and $\frac{1023}{1024}$, or even further (Van Brummelen 2009, pp. 141–142, 236–238).

1.1.2 Al-Kāshī and Iterative Approximations for the Sine of 1°

Nearly 1300 years after Ptolemy, the renowned Iranian mathematician-astronomer Ghiyāth al-Dīn Jamshīd al-Kāshī (d. 1429), in his Persian *zīj* or astronomical table-text of 1413/1414 entitled *Zīj-i Khāqānī*, continued the minor tradition of refining Ptolemy's double-inequality approximation. Al-Kāshī was a leading member of the scientific team at the Samarqand observa-

[2] To represent in this work the sexagesimal or base-60 place-value numbers common to Hellenistic, Islamic and Indian trigonometry, we follow the standard practice of using commas to separate the individual sexagesimal "digits", and a semicolon to separate the integer and fractional parts of a sexagesimal number. Thus, for example, the number $a, b; c, d$ represents $a \cdot 60 + b + c/60 + d/60^2$. These sexagesimal places are often designated by the familiar arclength/angle units "degrees", "minutes", "seconds", even when the quantity in question is not an arc or angle. Fractional places beyond seconds are denoted by their ordinal numbers, such as "thirds", "fourths", etc.

tory presided over by the Timurid prince and polymath Ulugh Beg (Mīrzā Muḥammad Ṭaraghāy, 1394–1449). Using the sines of $\frac{9}{8}^{\circ}$ and $\frac{15}{16}^{\circ}$, al-Kāshī was able to obtain $R\sin 1^{\circ} \approx 1;2,49,43,13,5$ (as opposed to the accurate value 1;2,49,43,11,15).[3]

However, much more important was al-Kāshī's devising of a revolutionary new approach for deriving the sine of 1° to arbitrary accuracy, shifting the methodological focus from geometry to algebra. This achievement appears to be first attested in a series of marginal glosses in a manuscript of his *Zīj-i Khāqānī* regarding his improved "Ptolemaic" value. The marginal notes, evidently transcribed after the deaths of both al-Kāshī and Ulugh Beg but including remarks attributed originally to al-Kāshī himself, allude to his new method for computing the same quantity to nine sexagesimal places of the mantissa:[4]

> The ancients found no method for obtaining accurately the sine of the third of an angle whose sine is known. We have developed a method and have written an essay explaining it to its limit. We have extracted the sine of one degree by that method; it is 1,2,49,43,11,14,44,16,19,16 ninths. *This is an annotation of the author; may God forgive him.*
>
> Mostly (?) the successful completion of this essay was not attained. After the death of the author, *upon him (be) forgiveness!*, this weak person (i.e., the present scribe) saw his book in which he had commanded the extraction (i.e., he had extracted) this sine of one degree by the method of algebra without question…(word illegible) and on its margin a writing that *'This is the sine of one degree; we extracted it by inspired strength from the Eternal Presence'.*
>
> *Our master,* Muᶜīn ud-Dīn, *the (astronomical) observer.*
>
> This indigent destitute person Muḥammad…(word illegible) after the martyrdom of the martyred Sultan Ulugh Beg (saw, or found?) in the portfolio of that presence (i.e., honorable person) a scrap (of paper?) in the handwriting of our forgiven (i.e., deceased) master.

There is no known surviving exposition of this "method of algebra" in al-Kāshī's own writings, but several of his successors provided elaborations and demonstrations of it (Rosenfeld and Hogendijk 2003, p. 27). The technique involves the iterative solution of a cubic equation derived from identities, the most important of which is an equivalent to the sine triple-angle formula, in terms of the known quantity $R\sin 3^{\circ}$ (or Crd 6°). Al-Kāshī's cubic equation, whose root is $R\sin 1^{\circ}$ (or Crd 2°), is of the form designated in the 11th-century algebra of ᶜUmar Khayyām as "a cube and a number are equal to things" (Kasir 1972, pp. 76–78). In modern parlance, this means a variable is equal to a multiple of its cube plus a constant, without any quadratic term, commonly called today a "depressed cubic" equation.

ᶜUmar Khayyām had solved cubic equations of this type only geometrically, by translating the coefficients into lengths and using conic sections to construct the unknown quantity as a line segment of the required length. Since the analytic solution to al-Kāshī's depressed cubic was not known in his own time, his solution relied instead upon a digit-recurrence algorithm reminiscent of currently familiar ones for long division and (less familiarly) root extraction. In particular, the modern sine triple-angle formula

$$\sin(\alpha) = 3\sin\left(\frac{\alpha}{3}\right) - 4\sin^3\left(\frac{\alpha}{3}\right), \quad \text{or more generally} \quad R\sin(\alpha) = 3R\sin\left(\frac{\alpha}{3}\right) - \frac{4\left(R\sin\left(\frac{\alpha}{3}\right)\right)^3}{R^2}, \tag{1.3}$$

is itself a depressed cubic equation, assuming that $\sin(\alpha)$ is known and $\sin\left(\frac{\alpha}{3}\right)$ is not. If we set $\alpha = 3^{\circ}$, then we have the cubic equation that we need to solve to find the sine of 1°. It applies either to $R\sin 1^{\circ}$ itself, which we denote x throughout this work, or to its double Crd 2°, which we denote y. If we assume $R = 60$ or $1,0$ sexagesimally, we can write canonical forms of the cubic equation for x or y as follows:

$$\begin{aligned} x &= \frac{(15,0)\cdot R\sin 3^{\circ}}{45,0} + \frac{x^3}{45,0}, \\ y &= \frac{(1,0,0)\cdot \mathrm{Crd}\,6^{\circ}}{3,0,0} + \frac{y^3}{3,0,0}. \end{aligned} \tag{1.4}$$

Al-Kāshī's approach to deriving these quantities, like other digit-recurrence algorithms in basic arithmetic, begins by extracting the leading sexagesimal digit of the solution, and generates additional digits of its fractional part one at a time. Throughout this book we call this type of solution technique "successive-digits iteration".

A crucial text in the dissemination of al-Kāshī's discovery was the Persian *Zīj-i Jadīd-i Sulṭānī* (or *Zīj-i Ulugh Beg*), composed in the second quarter of the 15th century in the Samarqand observatory. The impact of this work and its commentaries—especially those of Mīram Čelebī (1475–1525) and ᶜAbd ᶜAlī al-Bīrjandī (d. 1525/6)—is explored further in chapter 4. In particular, by the early 16th century, some forms of iterative solution algorithm(s) had been introduced that bypassed the digit-recurrence approach. In these versions, each iteration computes a new approximate value of the entire unknown quantity rather than just its next sexagesimal digit. While these methods are simpler to describe, each iterative step takes more effort to perform. We call these approaches, which include many variant formulas with mostly trivial differences, "successive-values iteration".

---

[3] Chapter 2 section 1, "On Determining the Sine and the Versed Sine" (ff. 28v–32r), especially rule 6 (ff. 31r–32r); see (Van Brummelen n.d.).

[4] This translation of the notes in India Office MS 430 Ethé 2232, f. 32r is copied from (Aaboe 1954, pp. 28–29) and preserves that translator's convention of italicizing the Arabic insertions among the Persian sentences.

Typical iterative equations of this nature for the quantities $x = R\sin 1°$ and $y = \mathrm{Crd}\,2°$ (again assuming $R = 60$) can be represented as follows:

$$\left.\begin{aligned} x_{k+1} &= x_0 + (0;0,1,20)\,(x_k)^3, \qquad & x_0 &= \frac{R\sin 3°}{3}, \\ y_{k+1} &= y_0 + (0;0,0,20)\,(y_k)^3, \qquad & y_0 &= \frac{\mathrm{Crd}\,6°}{3}, \end{aligned}\right\} \quad k \in \{0, 1, 2, 3, \ldots\}. \tag{1.5}$$

This "successive-values" approach is structurally similar to other fixed-point iteration techniques sometimes employed in Indian trigonometric computations since the mid-first millennium.

European astronomers followed similar paths in their trigonometric calculations, albeit a century or two later. Methods of computing sine tables before the 15th century relied on techniques that had been developed in Muslim Spain, based originally on techniques in Ptolemy's *Almagest*. Around the time of al-Kāshī, the Italian astronomer Giovanni Bianchini computed $R\sin 20°$ (which can be used together with the known $R\sin 21°$ to find $R\sin 1°$ analytically), using an informal but effective method of successive guesses (Gerl 1989, pp. 265–268). In the late 16th century, François Viète proposed finding small sine values by solving the sine triple-angle formula numerically, as al-Kāshī had done. Viète went as far as prescribing the determination of the sine of a single arcminute from that of a single degree, using not only angle trisection via the triple-angle formula, but also angle quintisection via the fifth-degree sine quintuple-angle formula. Although he did not implement his ideas with trigonometric tables, his methods were used in the early 17th century by Henry Briggs (Van Brummelen 2021, pp. 22–28).

1.2 Evolution of Techniques for Calculating the Sine of 1° in Sanskrit Mathematics

Beginning in the early Common Era apparently with Indian adaptation of the Hellenistic chord concept into the more versatile sine, Indian trigonometry developed along its own lines. Authors of trigonometric tables in the earliest extant Indian mathematical-astronomy treatises (Sanskrit *siddhāntas*) did not attempt to tabulate R sines of every degree of arc. Rather, they turned to geometry to determine a convenient tabular argument interval, based on the fact that the R sine of 30° is simply $R/2$. Beginning with $R/2$ and applying the sine half-angle formula three times in succession yields the R sine of 3°45′. This arc became the typical standard increment for tabulated arguments of Indian trigonometric functions; linear or superlinear interpolation algorithms were employed for intermediate values (Gupta 1969), (Hayashi 1997).

For the trigonometric radius R, Indian texts used a variety of different numbers rather than adopting a single standard value such as the canonical Greco-Islamic $R = 60$. Indian R numbers were usually integers large enough to allow for reasonably accurate integer representation of trigonometric function values. The most common radius was $R = 3438 \approx 360 \cdot 60/(2\pi)$, which might be thought of as a sort of "proto-radian" metrology, in that it makes the unit of radial length approximately equal to the arclength of one arcminute.

Since the content of astronomical *siddhāntas*—including numerical data such as R sine tables—was conventionally composed and transmitted in Sanskrit verse intended for memorization and oral recitation, authors and students alike appreciated the convenience of shorter, coarser tables combined with interpolation rules. Beginning as early as the seventh century, various rational approximations and iterative algorithms were devised to determine trigonometric quantities for arbitrary arcs, avoiding tabulated values and interpolation altogether (Hayashi 1991), (Plofker 2004). Such approximative methods continued to flourish alongside the ubiquitous versified tables of trigonometric values.

Within the Sanskrit textual tradition, efforts to attain an accurate standard value for the sine of 1° in particular go back at least to the *jyotpatti* or "sine-determination" chapter in Bhāskara II's renowned 1150 *Siddhāntaśiromaṇi* ("crest-jewel of *siddhāntas*") (Plofker 2009, pp. 204–205), (Montelle 2019). In the process of stating R sine sum and difference formulas, Bhāskara noted (without derivation or demonstration) that tabulated values of the R sine for arguments at intervals of a single degree can be produced by adding a given R sine value minus its own 6569th part to the corresponding R cosine times $10/573$ (Śāstrī 1989, pp. 282–285). This implies that $\sin 1° \approx 10/573$ and $\cos 1° \approx 6568/6569$, both accurate to the sixth decimal place.

About two and a half centuries after Bhāskara, the mathematician-astronomer Mādhava in the Kerala region of southwestern India discovered several power series for trigonometric functions, using infinitesimal arguments somewhat similar to those later developed in European calculus. Such series allowed the computation of any sine value to any desired accuracy (Gold and Pingree 1991), (Sarma et al. 2008), (Ramasubramanian and Sriram 2011). Mādhava's innovations were expanded with ingenious demonstrations and explanations by his successors, including some iterative techniques for approximating small-angle $R\sin$ values proposed in the early 15th century by Mādhava's pupil Parameśvara (Plofker 1996). The Mādhava school's discoveries, however, seem to have remained geographically isolated in their region of origin until the 19th century.

In the meantime, Indian and Greco-Islamic trigonometric traditions met again in the Indo-Islamic empires of second-millennium north India Scholars on both sides incorporated new concepts and technical vocabulary from translated texts into their own astronomy and astrology. One of the chief expositors of Islamic mathematical astronomy in Sanskrit was Nityānanda, an astronomer and protégé of the vizier Āṣaf Khān at the Mughal court of Shāh Jahān in Delhi in the early 17th century (Misra 2016). Nityānanda's 1639 *Sarvasiddhāntarāja* ("king of all *siddhāntas*") was a pioneering attempt "to systematically syncretise Islamic astronomy within the traditional *siddhāntic jyotiṣa* astronomy" (Misra 2016, p. iii). In particular, it includes the first known Sanskrit presentation of the iterative approaches to computing $R\sin 1°$ associated with the work of al-Kāshī and Ulugh Beg. Nityānanda's treatment of this problem is described in detail in (Montelle and Ramasubramanian 2018) and summarized in section 5.1 below. By contrast, the *Siddhāntatattvaviveka* ("investigation of the truth of *siddhāntas*") composed by Kamalākara in Varanasi in 1658 briefly alludes to Ulugh Beg's trigonometry, but does not discuss it in detail (Dvivedi and Jha 1991, p. 138).

1.2.1 The Sine of 1° in 18th-Century Jaipur

Both Nityānanda's *Sarvasiddhāntarāja* and Kamalākara's *Siddhāntatattvaviveka* influenced the work of Jagannātha Samrāṭ, one of the most prominent scholars at the court of another astronomer-prince, Jayasiṃha (Jai Singh II, 1688–1743) of the princely state of Amber or Jaipur in Rajasthan.[5] Erudite pandits in various genres of Sanskrit scientific literature, Jagannātha and his colleague Nayanasukha Upādhyāya were also active in Jayasiṃha's ambitious program in mathematics and astronomy, which encompassed not only the composition of many Sanskrit works but also the study and translation of Arabic and Persian texts. This program eventually grew to include the investigation of European books in Latin, with scholarly collaborations involving Hindu, Muslim and European Jesuit colleagues.

Jagannātha is known for his Sanskrit translations of Naṣīr al-Dīn al-Ṭūsī's versions of Euclid's *Elements* (Sanskrit *Rekhāgaṇita* "line-computation, geometry", 1727) and Ptolemy's *Almagest* (Sanskrit *Siddhāntasamrāṭ* or *Samrāṭsiddhānta* "lord of treatises", 1732?). His labors at the Jaipur court included the composition, around 1727, of a verse work called *Siddhāntakaustubha* ("divine jewel of *siddhāntas*"), whose section on trigonometry in part echoes and expands the $R\sin 1°$ computation methods discussed earlier by Nityānanda. At some point (mid-1730's?) the *Siddhāntakaustubha*, enriched by sections of prose commentary or addenda composed perhaps by Jagannātha himself or by his colleague Nayanasukha, was incorporated into the *Siddhāntasamrāṭ* in the form of several supplementary chapters.[6]

We have centered this study upon a part of one *Siddhāntakaustubha*/*Siddhāntasamrāṭ* section containing an extensive discussion of trigonometry, including several iterative methods for determining $R\sin 1°$. In its expanded version included in the *Siddhāntasamrāṭ*, this section is headed as follows (Sharma 1967–1969, p. 1048):

atha vakṣyamāṇagaṇitopayogijyācāpādikaṃ rekhāgaṇitasādhyaṃ pratipādyate

Now begins the geometric determination [relating to] the sine and arc (*jyā-cāpa*), etc., employing the subsequently stated calculation.

We refer to this accreted text henceforth as the *jyācāpa* section of the *Siddhāntakaustubha* or *Siddhāntasamrāṭ*.

The Sanskrit prose addenda in the *jyācāpa* section have been partially preserved in draft form, with some intriguing variations, in manuscript 269:I.13 i 3 of the Sri Ram Charan Museum of Indology in Jaipur (Sharma 1986–1994, p. I.77), which as far as we know has not been previously studied. The anonymous manuscript combines two slightly differing copies or recensions of the textual material, some of which explicitly draws upon the early 16th-century Persian commentary or *Sharḥ* of al-Bīrjandī on the *Zīj-i Ulugh Beg* for its methods of computing the sine of 1°. It also describes variations on such methods attributed to two known Muslim scientists (*nujūmīs*, practitioners of *ʿilm al-nujūm* or astronomy) at Jayasiṃha's court: Muḥammad Ābida or Ābidda, who worked with Nayanasukha on other Persian-Sanskrit scientific translation projects, and Dayānat Khān.[7]

These expositions by Nayanasukha and/or Jagannātha, evidently based on deep study and collaboration with their Muslim colleagues and other scholars, furnish the latest known contribution to a textual tradition of remarkable breadth and complexity

[5] See especially (Pingree 2004), (Sharma 1995, p. 257). A comprehensive history of the multi-lingual and multi-disciplinary scholarly activities of Jayasiṃha's court scientists is beyond the scope of this work. For further details, including information about the two Muslim scientists named later in this section, the reader is referred also to (Mercier 1993), (Sharma 1995), (Pingree 1999), (Pingree 2003), (Kusuba and Pingree 2002), and the many related works referenced in them.

[6] For the identification of the work(s) in question, see (Sharma 1995, pp. 274–275) and (Pingree 2004, S1–S5). The full textual history and even the title(s) connected with the *Siddhāntakaustubha*/*Siddhāntasamrāṭ* remain somewhat obscure (Sharma 1995, pp. 258–261). The *Siddhāntakaustubha* has been published as a standalone verse text in (Pingree 2004), which we abbreviate *SK*, and its *Siddhāntasamrāṭ*-supplement version in (Sharma 1967–1969), J_S for short. A different form of the *Siddhāntasamrāṭ* version was published in (Chaturveda 1976) according to (Pingree 2004, S4–S5), but we have not yet been able to consult it.

[7] Ābida is likely identical with the *nujūmī* named in Jaipur court records as Muḥammad Ābid Mullā, while Dayānat Khān was Jayasiṃha's "most favored and decorated Muslim astronomer" (Sharma 1995, pp. 282–283).

on the sine 1° problem. Their synthesis of various approaches shows a profound understanding of the concepts and methods involved, as well as providing insights into little-known details of mathematical practice. Perhaps most importantly, their textual recensions constitute a rare example of a direct Sanskrit record of interaction between Sanskrit and Persian/Arabic scientific traditions and their distinct approaches to mathematical topics including trigonometry, algebra, and sexagesimal arithmetic. The mutual influences of these approaches in the early modern period are still only beginning to be understood.

1.3 The *jyācāpa* Excerpt: Text, Manuscript and Editing

The *jyācāpa* prose portions preserved in the Jaipur manuscript SRCMI 269:I.13 i 3 are in part comments upon a set of about two dozen Sanskrit verses that seem to have appeared initially in the verse-only recension of Jagannātha's *Siddhāntakaustubha*, and subsequently in slightly modified form in the expanded *Siddhāntasamrāṭ*. The manuscript is catalogued under the title *rekhā gaṇita (phiraṃgī matānusāra)* or "geometry (according to European views)", with the heading *jyācāpopapatti* ("demonstration [involving] sine and arc") on the first page.

The manuscript's two separately foliated partial drafts of *jyācāpa* material are here denoted J_1, which occupies the first ten of its 16 folia, and J_2, occupying the last six. Table 1.1 collates the sections of their textual content with corresponding material in published versions of the *Siddhāntakaustubha* (*SK*) and *Siddhāntasamrāṭ* (J_S).

Lacking a colophon or any explicit statement of the work's date, provenance, title, author or scribe, the manuscript's 16 leaves are written in portrait orientation in clear *nāgarī* script with ochre highlighting, with dimensions about 16 × 23 cm. The two parts J_1 and J_2, written in different hands, each contain several Euclidean-style labeled diagrams with *nāgarī* letter labels, as well as some displayed equations marked off from the text in boxes. The chief individual characteristics of the two parts are as follows (see facsimile examples in figures 1.2 and 1.3):

J_1 Ff. 1–10, f. 10v blank. Foliated in lower left corner of verso. 18–24 lines per page. A few marginal corrections. Large blank area (for diagram?) on f. 4v.

J_2 Ff. 1–6. Foliated in upper left corner of verso. 29–36 lines per page. Occasional marginal corrections. Blank area (for diagram?) on f. 5r. F. 6v consists of a large computation grid.

The following conventions are observed in the edited and translated Sanskrit text:

- The text in manuscript part J_1 is edited and translated in the left column, and that of part J_2 in the right column, with paragraph breaks added for readability. Missing text in either column denotes its absence in that recension. The critical apparatus for each text is given in footnotes.
- Long vertical strokes in the edited text, accompanied by square-bracketed folio/page number identifiers, indicate the occurrence of folio breaks in the manuscript or page breaks in the printed edition J_S (Sharma 1967–1969). (Although we have occasionally consulted the edited text or variant readings recorded in J_S to check uncertain readings in our manuscript, we have not collated them systematically with our edition.)
- A thick vertical bar (**|**) in the edition represents an editorially inserted single *daṇḍa*. (The *daṇḍa* is the Sanskrit punctuation mark, consisting of single or double vertical strokes.)
- Scribal variants of sandhi or *nāgarī* orthography which are usually emended silently without mention in the critical apparatus (except where the meaning of the original reading may be ambiguous) include the following: *anusvāra* for a conjoined nasal, omitted *visarga*, *virāma* or *avagraha*, reversed conjunct consonants (e.g., *adba* for *abda*), conjunct consonants that we cannot reproduce in our *nāgarī* typesetting, doubled consonants after *r* or across a *pāda* break, and routinely conflated consonant pairs (e.g., *ba* for *va*, *ṣa* for *kha*).
- We have not regularized the scribes' sandhi in the case of mathematical notation (particularly in the representation of numbers and letter labels in diagrams), technical terms, or transliterations from Persian. We have also deliberately retained in the translation the occasional awkwardness of Sanskrit syntax accompanying the author's translation from Persian.
- Square brackets in the translation denote editorial additions. In particular, we have supplied for the reader's convenience the section breaks and section titles in square brackets, following topic shifts indicated by ochre pigment highlighting in the manuscript.
- The Sanskrit overdot symbol indicating a negative number is represented in the translation by the conventional minus sign.
- English abbreviations and notation conventions (e.g., the familiar symbols for degrees and arcminutes) are used to translate Sanskrit syllabic abbreviations (which typically are signified by the standard *nāgarī* elision symbol of an open circle). However, we have transliterated instead of translating the following standard Sanskrit technical terms and abbreviations:

paṅkti "row", sequence of successive digits extracted by, e.g., a division or root algorithm
rūpa (*rū*) "form", constant term, unit
yāvattāvat (*yā*) "as much as so much", unknown quantity (first power of variable), or its coefficient

Textual content	*SK*	J_S	J_1	J_2
jyotpatti "sine-determination"	vv. II.1.48–59, pp. S43–S45	vv. 1–12, pp. 1048–1050	—	—
dviguṇāṃśajīvā "sine of twice the degrees"	vv. II.1.60–61, p. S46	v. 13, p. 1050	—	—
cāpajyājñāne sati cāpatrtīyāṃśajyānayana "derivation of the sine of one-third the arc when the sine of the arc is known"	vv. II.1.62–71, pp. S46–S47	vv. 14–23, pp. 1050–1051	—	—
cāpajñāne sati pañcamāṃśajyānayana "derivation of the sine of one-fifth the arc when the arc is known"	vv. II.1.72–76, p. S48	vv. 24–25, p. 1051	—	—
commentary on *jyotpatti* and *dviguṇāṃśajīvā* sections	—	pp. 1051–1058	—	—
citation of al-Bīrjandī on al-Kāshī and Ulugh Beg	—	—	—	f. 1r
commentary on *SKSS* vv. 14–18, on methods 2.1 and 2.2	—	pp. 1058–1059	ff. 1r–1v	—
beginning of demonstration for method 2.3	—	pp. 1059–1060	f. 1v	f. 1r
demonstration of *Almagest* 1.4	—	pp. 1060–1061	ff. 1v–2r	—
continuation of demonstration for method 2.3	—	pp. 1061–1067	ff. 2r–5v	ff. 1r–3r
method 2.4, attributed to Ābida	—	pp. 1067–1068	ff. 5v–6r	f. 3r
method 2.5, attributed to Ulugh Beg	—	pp. 1068–1070	ff. 6r–7v	ff. 3r–4r
method 2.6, also attributed to Ulugh Beg	—	pp. 1070–1073	ff. 7v–9r	ff. 4r–4v
alternate demonstration 2.7 for method 2.6	—	pp. 1073–1074	ff. 9r–10r	ff. 4v–5r
method 2.8, attributed to Dayānat Khān	—	—	—	ff. 5r–6v
abridgement of method 2.8	—	pp. 1074–1076	—	—
additional material on sines	v. II.1.77, and table; pp. S49–S51	—	—	—
additional material on sines	—	vv. 1–25, and commentary; pp. 1076–1085	—	—

Table 1.1: Portions of the *jyācāpa* material as they appear in the edition of the original verse recension of the *Siddhāntakaustubha* (*SK*), (Pingree 2004); in its expanded form in the published *Siddhāntasamrāṭ* (J_S), (Sharma 1967–1969); and/or in parts J_1 and J_2 of the Jaipur manuscript. The numbers we assigned to the different methods correspond to the sections where they are discussed in chapter 2.

yāvattāvat-varga (*yāva*) square of unknown quantity, or its coefficient
yāvattāvat-ghana (*yāgha*) cube of unknown quantity, or its coefficient
yāvattāvat-varga-varga (*yāvava*) fourth power of unknown quantity, or its coefficient

The problem of how to deal with numerical inaccuracies in the surviving text has been extraordinarily vexing, and the reader will probably not be much more fully satisfied with our chosen solution than we are. It seems to us that the following approach is the least misleading way to represent the introduction of an "incorrect" number:

- If it appears to be only a trivial scribal error that does not propagate through subsequent computational steps (especially if it is confined to only one of the two manuscript versions), we emend it like any other textual variant with a note in the apparatus.
- If it is a more obviously substantial mistake that affects the dependent computations in some way—whether the error originated in one of the source texts, or through the author's misreading of an Arabic/Persian alphabetic (*abjad*) number in an Islamic source text, or in the author's own computations—we write the incorrect number(s) in red in both text and translation. However, subsequent errors that may have been caused by the use of this incorrect number in the computations immediately following it are *not* so flagged.
- A full reconstruction of each calculation in the textual material analyzed in chapters 3, 4 and 5 has been appended to the corresponding mathematical commentary, including brief descriptions of any numerical anomalies and their consequences.

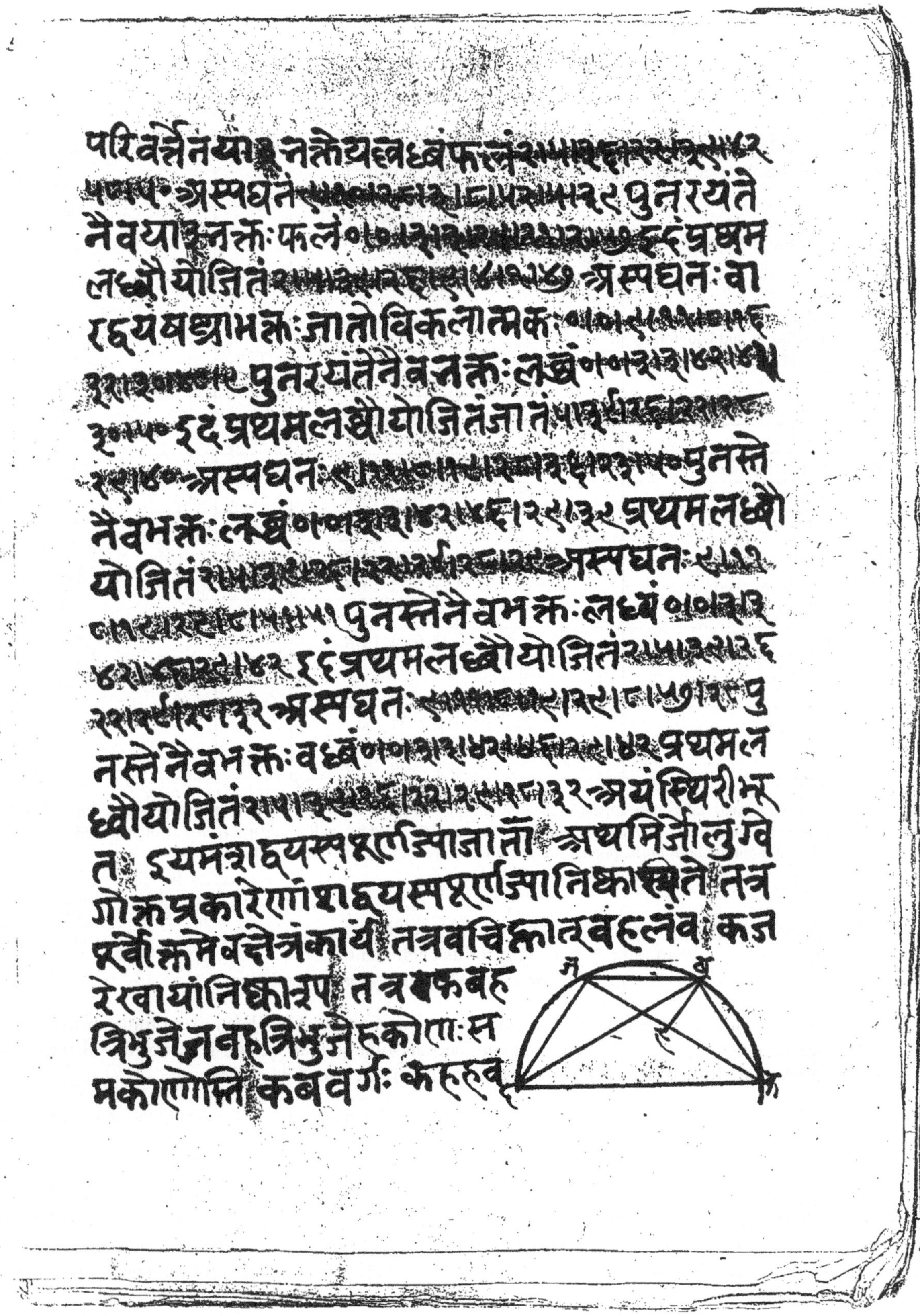

Figure 1.2: Facsimile of f. 6r in part J_1 of MS SRCMI 269:I.13 i 3 showing text, sexagesimal numerical values, and a geometric diagram. A value for the chord of 2° (2; 5, 39, 26, 22, 29, 28, 32) appears seven lines from the bottom.

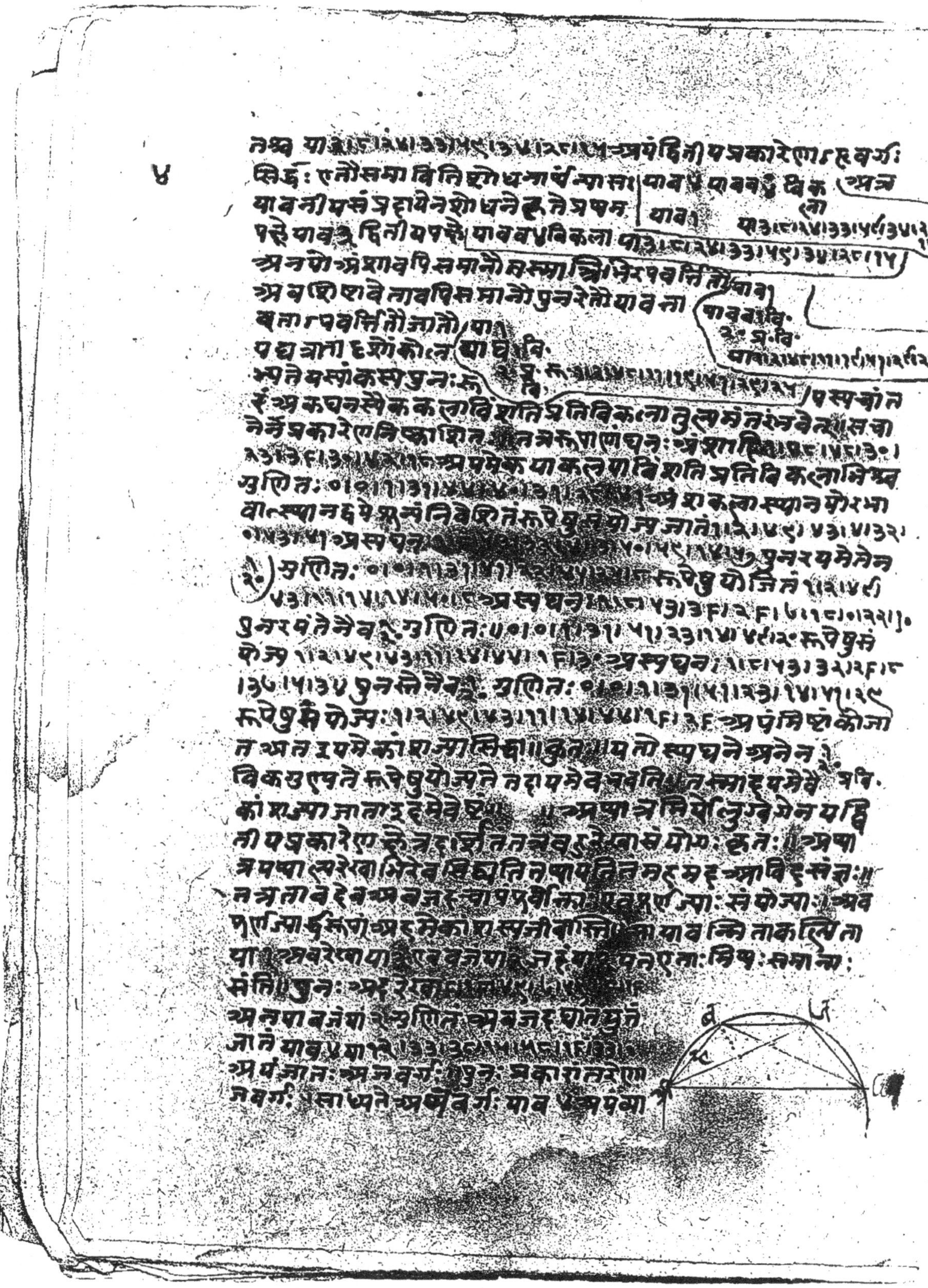

Figure 1.3: Facsimile of f. 4v in part J_2 of MS SRCMI 269:I.13 i 3 showing text, numerical values, a geometric diagram, and displayed equations (top right). A value for the sine of 1° (1; 2, 49, 53, 11, 14, 44, 16, 26) appears 14 lines from the bottom.

Chapter 2
Parallel Editions and Translations of the Manuscript Text Recensions of the *jyācāpa* excerpt

|[f. 1r J_1] ॥ श्रीगणेशाय नमः ॥

|[p. 1058 J_S] अथांशत्रयस्य जीवाज्ञाने सत्येकांशजीवाज्ञानं दोर्ज्यामिति श्लोकचतुष्टयेनाह ॥

Homage to Lord Gaṇeśa.

Now, when there is knowledge of the sine of three degrees, he [i.e., Jagannātha Samrāṭ] states the knowledge of the sine of one degree with four verses [beginning] "The sine of the arc".

|[f. 1r J_2] ॥ श्रीगणेशाय नमः ॥

अथैकांशजीवाविषये उलुग्वेगीजीकस्य शरहविर्जंदीस्थव्याख्या लिख्यते ॥ तत्रैकांशजीवानयने नव्यतरं प्रकारद्वयमस्ति ॥ एकं यमशैदकाशीसंज्ञेन कृतम् ॥ द्वितीयं मिर्योलुग्वेगेन कृतम् ॥ परं चोलुग्वेगस्य यमसैदकाशीवेधप्रक्रियायां साहाय्यकार्यस्ति[1] ।

Homage to Lord Gaṇeśa.

Now, on the subject of the sine of one degree, the commentary in the *Śarahavirjaṃdī* [*Sharḥ-i Zīj-i Jadīd-i Sulṭānī* of al-Bīrjandī] of the *Ulugvegī zīj* [*Zīj-i Ulugh Beg*] is written. In it there are a couple of newer methods on the determination of the sine of one degree. One is made by [the one] named Yamaśaida Kāśī [Jamshīd al-Kāshī]; the second is made by Miryolugvega [Mīrzā Ulugh Beg]. And the later [work?] of Ulugveg is helpful in [understanding?] the breakthrough method of Yamasaida Kāśī.

[1] साहाथकार्यस्ति J_2

C. Montelle et al., *From Samarqand to Jaipur*, Sources and Studies in the History of Mathematics and Physical Sciences, https://doi.org/10.1007/978-3-031-89087-1_2

2.1 [Successive-Digits Iteration for Crd$(2\alpha/3)$ and $R\sin(\alpha/3)$]

करसङ्गुणितां द्विगुणां भुजज्यां त्रिभक्तां विधाय गणकप्रवीणः फलमंशास्तान्पृथक् स्थापयेत् ॥ तु परं तेषां फलरूपांशानां घने व्यासार्धवर्गभक्ते सति |[p. 1059 J*s*] यत्फलं तद्भागावशेषसहितं कृत्वा त्रिभक्तं कार्यं ॥ त्रिभक्ते यत्फलं कलात्मकं तत्पूर्वलब्धांशरूपफलाधः स्थाप्यं सांशकलात्मिका पङ्क्तिर्जाता ॥

पङ्क्तेर्घनः प्रथमघनेन हीनो व्यासार्धवर्गभक्तः। फलं शेषे योज्यं[1] त्रिभिर्भजेत् ॥ लब्धं फलं विकलाः कलाधः स्थाप्याः। सांशकलाविकलात्मिका पङ्क्तिर्जाता ॥ पङ्क्तेर्घनः कार्यः। अस्मिन्पूर्वघनः शोध्यः। शेषे त्रिभज्यावर्गभक्ते यत्फलं तद्भाज्यावशेषसहितं कृत्वा मुहुर्वारं वारं तदग्रे तु परं पुरोक्तो विधिः कार्यः। एवं कृते फलं पङ्क्तिः स्यात् ॥ पङ्क्त्यर्धं चापस्य तृतीयांशज्या स्यात् ॥

"Having determined the sine" of the arc [and made it] "*kara*-multiplied", [i.e.,] multiplied by two, and "divided by three", the skillful one among mathematicians should set those down separately (the result [is] degrees). But when the cube of those, [i.e.,] of the degrees of the *rūpa* of the result, is divided by the square of the half-diameter: whatever is the result, [making] that "increased by the remainder from [dropping] the degrees", [it] is to be divided by three. When divided by three, whatever is the result in minutes, that is to be set down below the result from the *rūpa* of the degrees of the previous quotient; that is the *paṅkti* produced in degrees and minutes.

The cube of the *paṅkti* diminished by the first cube [is] divided by the square of the half-diameter. One should divide the result, [which is] to be added to the remainder, by three. The resulting quotient in seconds [is] to be set below the minutes. That resulting *paṅkti* [has] the form degrees, minutes, seconds. The cube of the *paṅkti* is to be made. The cube of the previous [*paṅkti*] is to be subtracted from that. When the remainder is divided by the square of the radius, whatever is the result, [when one has] increased that by "the remainder from the division", the "previously-stated" procedure is to be done "repeatedly", [i.e.,] again and again, "but at the end of that" [*paṅkti*]. When done thus, the result is the [desired] *paṅkti*. Half the *paṅkti* is the sine of a third part of the arc.

---

[1] योज्य J₁

2.2 [Successive-Values Iterations for Crd$(2\alpha/3)$ and $R\sin(\alpha/3)$]

अथेत्यनन्तरं तां त्रिभागजीवां पुनः प्रकाराद्वच्मि ॥ यतः कारणादनेकैर्भेदैर्गणकस्य बुद्धिः स्वशास्त्रे प्रवीणतामेति ॥ १६ ॥

पुनः प्रकारान्तरेण त्रिभागजीवामाह ॥ स्वत्र्यंशहीनेति ॥ भुजज्या स्वतृतीयांशेन हीना पृथक्स्थाप्या ॥ अस्या[2] घनस्त्रिगुणस्य कृत्या त्रिघ्न्या भक्तः। लब्धफलेन पृथक्स्था भुजज्या युता कार्या। मुहुरनया क्रियया पङ्क्तिः साध्या। पङ्क्त्यर्धं चापगुणांशजीवा धनुस्त्र्यंशज्या स्यात् ॥ १७ ॥

पुनः प्रकारान्तरेणाह ॥ भुजांशजीवेति ॥ [f. 1v J_1] भु|जांशज्यायास्तृतीयांशः पृथक्स्थाप्यः ॥ तस्य घने स्वांशयुते त्रिज्यावर्गभक्ते लब्धं यत्फलं तेन पृथक्स्थो युतः कार्यः ॥ एवं मुहुर्वारं वारमनया क्रियया धनुस्त्रिभागजीवा स्यात् ॥ १८ ॥

"Now": immediately, "I state that", the sine of a third part, "furthermore according to [another] method". "Since" (from which cause) "the wisdom of the mathematician attains skill in his own discipline by means of many variations." (16)

"Furthermore" by another method he states the sine of a third part [of the arc]. "Diminished by its own third part": The sine of the [given] arc, diminished by its own third part, is to be set down separately. The cube of it is divided "by three times the square of the radius". The [diminished] sine of the arc standing separately is to be increased by the resulting quotient. "Repeatedly" with that procedure a *paṅkti* is to be determined. Half the *paṅkti* is "the sine of a third part of the arc", the sine of a third part of the arc. (17)

Furthermore, he states by another method: "The sine of the degrees of the arc". A third part of the sine of the degrees of the arc is to be set down separately. When its cube is added to its own part [i.e., its own one-third], [and] divided by the square of the half-diameter: whatever is the resulting quotient, the [quantity] standing separately is to be increased by that. In this way "repeatedly", again and again, with that procedure the sine of a third part of the arc is [computed]. (18)

[2] अस्याः J_1

2.3 [Al-Kāshī's Successive-Digits Iteration for the Chord of 2°]

अस्योपपत्तिः ॥ सा यथा ॥ कबजदषडंशचापं कल्पितम् ॥ पुनरस्य समानं भागत्रयं कल्पितं बचिह्नजचिह्नयोः ॥ तत्र कबं कजं कदं बजं बदं जदं चैताः पूर्णज्यारेखा योज्याः ॥

अथात्र पूर्वोक्तप्रकारेणांशत्रयस्य जीवा ज्ञातास्ति। सा चेयं ३। ८। २४। ३३। ५९। ३४। २८। १४। ५०। इयं द्विगुणा ६। १६। ४९। ७। ५९। ८। ५६। २९। ४० जाता कदपूर्णज्या ॥

अत्रांशद्वयस्य कबपूर्णज्याज्ञानमिष्टमस्ति ॥ अस्योपपत्त्यर्थं[3] किंचिदुच्यते ॥ तत्र |[p. 1060 J_S] सिद्धान्तसम्राजि प्रथमाध्यायस्य द्वितीयक्षेत्र इदमुपपादितम् ॥ यच्चतुर्भुजं वृत्तान्तः पतति तत्सन्मुखस्थभुजानां घातयोगस्तच्चतुर्भुजान्तः पतितकर्णद्वयघातसमो[4] भवतीत्यस्योपपत्तिस्तु पूर्वं कथितैवास्ति ॥

Its rationale: it is as [follows]. The six-degree arc *KBJD* is considered. Furthermore, three equal parts of it are considered at points *B* [and] *J*. Then these [segments] *KB*, *KJ*, *KD*, *BJ*, *BD*, and *JD* are to be joined [as] chord lines.

Now here by the previously stated method the sine of three degrees is known. And it is this: 3; 8, 24, 33, 59, 34, 28, 14, 50. This is multiplied by two: 6; 16, 49, 7, 59, 8, 56, 29, 40 is the resulting chord *KD*.

Here the knowledge of the chord *KB* of two degrees is desired. Something is said for the sake of its rationale. There in the *Siddhāntasamrāṭ*, in the second figure of the first chapter this is demonstrated. [See figure J_1.2.] "Whatever quadrilateral falls inside a circle, the sum of the products of the sides standing opposite becomes equal to the product of the two diagonals falling inside that quadrilateral"; but its rationale is just explained previously.

3 ०पत्य० J_1
4 पातिकर्ण० J_1

तत्र प्रथमप्रकारः कथ्यते। स[2] यथा ॥ अबजदं षडंश६चापं कल्पितम् ॥ पुनरस्य समानं भागत्रयं बचिह्नजचिह्नयोः कृतम् ॥ तत्र अबं अजं अदं बजं बदं जदं चैताः पूर्णज्यारेखा योज्याः ॥

अथात्र पूर्वोक्तप्रकारेणांशत्रयस्य[3] जीवा ज्ञातास्ति। सा चेयं ३। ८। २४। ३३। ५९। ३४। २८। १४। ५०[4]। इयं द्विगुणा ६। १६। ४९। ७। ५९। ८। ५६। २९। ४० जाता अदपूर्णज्या[5] ॥

अत्रांशद्वयस्य अबपूर्णज्याज्ञानमिष्टमस्ति ॥ तत्र मिजिस्तीग्रन्थस्य प्रथमाध्यायस्थद्वितीयक्षेत्र इदमुपपन्नम्[6] ॥ यच्चतुर्भुजं वृत्तान्तः पतति तत्सन्मुखस्थभुजानां[7] घातयोगस्तच्चतुर्भुजान्तः पतितकर्णद्वयघातसमो भवति ॥

Then the first method is stated. It is as [follows]: *ABJD* is considered an arc of six (6) degrees. Furthermore, three equal parts of it are made at points *B* and *J*. Then these [segments] *AB*, *AJ*, *AD*, *BJ*, *BD*, and *JD* are to be joined [as] chord lines.

Now here by the previously stated method the sine of three degrees is known. And it is this: 3; 8, 24, 33, 59, 34, 28, 14, 50. This is multiplied by two: 6; 16, 49, 7, 59, 8, 56, 29, 40 is the resulting chord *AD*.

Here the knowledge of the chord *AB* of two degrees is desired. There in the second figure in the first chapter of the book *Mijistī* this is explained. "Whatever quadrilateral falls inside a circle, the sum of the products of the sides standing opposite becomes equal to the product of the two diagonals falling inside that quadrilateral."

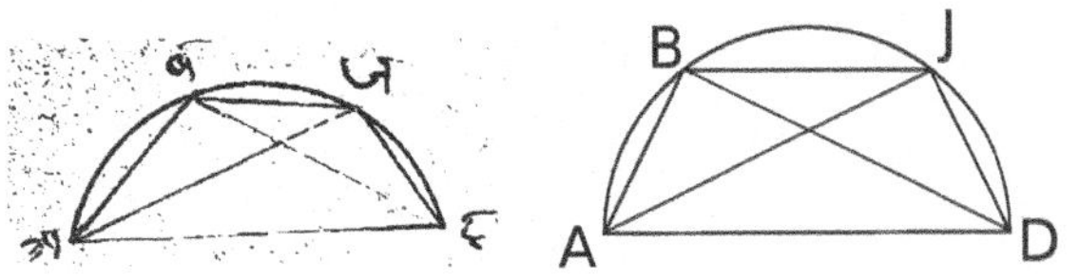

Figure J_2.1: Diagram on f. 1r of MS J_2 (facsimile and reconstruction)

2 सं J_2
3 ०पूर्वात्त० J_2
4 ०२८। ५४। ५० J_2
5 मदपर्णज्या J_2
6 क्षेत्र यिदमुपपन्नं J_2
7 तत्संन्मुख० J_2

पुनस्तत्रैव प्रथमाध्यायस्य चतुर्थक्षेत्र इदमुपपादितम् ॥ इष्टचापार्धपूर्णज्यावर्गस्तच्चापोनभार्धांशपूर्णज्याव्यासान्तरगुणितव्यासार्धतुल्यो भवति ॥

अस्योपपत्तिः ॥ कजव्यासे कबजदं वृत्तार्धं कार्यम् ॥ तत्र बदं दजं[5] सममस्ति ॥ कबरेखाकदरेखाबदरेखाजदरेखाः संयोज्याः ॥ दचिह्नात् दफरेखा कजव्यासे लम्बरूपा कार्या ॥ अत्र जफरेखा कजव्यासस्य ज्ञातचापोनभार्धांशपूर्णज्यायाश्चान्तरार्धमितास्ति ॥ कुतः ॥ कबं कहतुल्यं कार्यम् ॥ दहरेखा योज्या ॥ हकदत्रिभुजदकबत्रिभुजयोः। कहभुजकदभुजौ बकभुजक|दभुजयोः [f. 2r J_1] क्रमेण समानौ स्तः ॥ पुनः हकदकोणः बकदकोणेन समो ऽस्ति ॥ तस्मात् दहरेखा बदरेखा समाना जाता ॥ दजरेखापि समाना जाता ॥ तदा दजवर्गः दफवर्गजफवर्गयोर्योगेन तुल्यो भविष्यति ॥ एवं दहवर्गः दफवर्गफहवर्गयोगेन तुल्यो भविष्यति ॥ तस्मात् फहं जफं च समानं जातम् ॥ अनयोर्योगः कजकबयोरन्तरमस्ति ॥

Furthermore, just there in the fourth figure of the first chapter this is demonstrated. The square of the chord of half the desired arc is equal to the half-diameter multiplied by the difference of the diameter and the chord of the degrees of half a circle minus that arc.

Its rationale: On diameter KJ half-circle $KBJD$ is to be made. Then BD is equal [to] DJ. Line KB, line KD, line BD, [and] line JD are to be joined. From point D line DF is to be made in the form of a perpendicular on diameter KJ. Here line JF is measured by half the difference of diameter KJ and the chord of the degrees of half a circle diminished by the known arc. Why? KB is to be made equal to KH. Line DH is to be joined. In triangle HKD [and?] triangle DKB, side KH [and] side KD are respectively equal to side BK [and] side KD. Furthermore, angle HKD is equal to angle BKD. Therefore line DH is produced equal to line BD. Line DJ too is produced equal [to them]. Then the square of DJ will be equal to the sum of the square of DF [and] the square of JF. In the same way, the square of DH will be equal to the sum of the square of DF [and] the square of FH. From that, FH and JF [are] produced equal. The sum of those two is the difference of KJ and KB.

पुनस्तत्रैव प्रथमाध्यायस्य चतुर्थक्षेत्र इदमुपपन्नम् ॥ इष्टचापार्धपूर्णज्यावर्गस्तेनाङ्केन समो भवति ॥ यो ऽङ्कः[8]। तच्चापोनभार्धांशानां या पूर्णज्या भवति। अस्या व्यासेन यदन्तरं तेन गुणितम्[9]। यो व्यासार्धः। तदङ्कतुल्यो[10] भवति ॥

Furthermore, just there in the fourth figure of the first chapter this is explained. The square of the chord of half the desired arc is equal to that number. Which number [is]: Whatever is the chord of the degrees of half a circle minus that arc; whatever the difference of that with the diameter; whatever the half-diameter multiplied by that; is equal to that number.

[5] दजं inserted from left margin J_1

[8] यो कः J_2

[9] गुणित J_2

[10] व्यासार्धः स्तदंकतुल्यो J_2

पुनः कदजत्रिभुजे दसमकोणात् दफलम्बः कर्णोपर्यागतो ऽस्ति ॥ तस्मात्कजस्य निष्पत्तिर्जदेन तथास्ति यथा जदनिष्पत्तिर्जफेन ॥ तस्मा|[p. 1061 J_S] त्कजजफघातः दजवर्गेण समानो जातः ॥ तदेतदुपपन्नम्। इष्टचापार्धपूर्णज्यावर्गो चापोनभार्धांशपूर्णज्याव्यासान्तरगुणितव्यासार्धतुल्यो भवतीति ॥ शकलम् ॥

Furthermore, in triangle KDJ the perpendicular DF from the right angle D falls upon the hypotenuse. Therefore the ratio of KJ to JD is the same as the ratio of JD to JF. Therefore the product of KJ and JF [is] produced equal to the square of DJ. Just that is explained: "The square of the chord of half the desired arc is equal to the half-diameter multiplied by the difference of the diameter and the chord of the degrees of half a circle diminished by the arc." Figure:

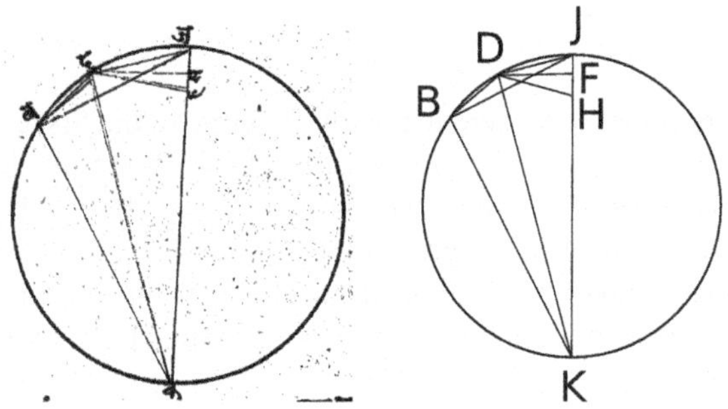

Figure J_1.1: Diagram on f. 2r of MS J_1 (facsimile and reconstruction); *Siddhāntasamrāṭ* ch. 1 fig. 5.

अथ प्रकृतमनुसरामः ॥ एवमत्रकथितक्षेत्रे कबचापजदचापे मिथः समाने स्तः ॥ एवं कजबदचापे च समाने स्तः ॥ तस्मात्कबजदरेखाघातो वर्गो भविष्यति ॥ अत्र बजमज्ञातं तद्यावत्तावन्मितं कल्पितम् ॥ इदं कदेन गुणितं जातं या० ६। १६। ४९। ७। ५९। ८। ५६। २९। ४०।

अस्य कबजदघातरूपयाव| [f. 2v J_1] द्वर्गस्य च योगः कजवर्गतुल्यकजबदघातेन समानो ऽस्ति ॥ अतो ऽयं कजवर्गतुल्यो जातस्तस्मात्कजवर्गराशावेको वर्गराशिरेभिर्यावत्तावद्भिरधिको जातः ॥ या०व० १ या० ६। १६। ४९। ७। ५९। ८। ५६। २९। ४० ॥ अस्य निदर्शनार्थं शकलं यथा ॥ ॥

Now, we pursue the [matter] under discussion. Thus here in the figure stated, arc KB [and] arc JD are equal to each other. And in the same way, arcs KJ and BD are equal. Therefore the product of lines KB [and] JD will be a square. Here, *BJ* [being] unknown, that is considered the amount of the *yāvattāvat*. This is multiplied by KD; result, *yā* 6; 16, 49, 7, 59, 8, 56, 29, 40.

The sum of it and the square of the *yāvattāvat* (a number [equal to] the product of KB [and] JD) is equal to the product of KJ [and] BD [which is] equal to the square of KJ. Hence this resulting [expression is] equal to the square of KJ; therefore in [place of] a quantity [equal to] the square of KJ, one square quantity increased by these [many] *yāvattāvats* [is] produced: *yāva* 1, *yā* 6; 16, 49, 7, 59, 8, 56, 29, 40. For the sake of showing that, a figure as [follows]:

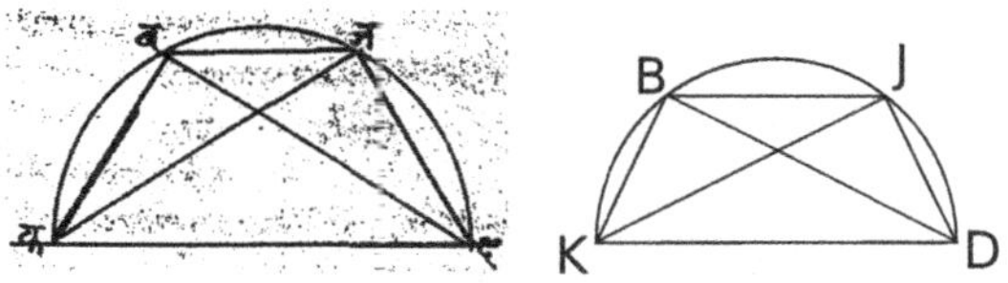

Figure J_1.2: Diagram on f. 2v of MS J_1 (facsimile and reconstruction); *Siddhāntasamrāṭ* ch. 1 fig. 4.

एवमत्र अबचापजदचापे मिथः समाने स्तः[11]। एवं अजबदचापे च समाने स्तः ॥ तस्मात् अबजदघातो वर्गो भविष्यति ॥ अत्र बजमज्ञातं तद्यावत्तावन्मितं कल्पितम्। इदमदेन गुणितं जातं या ६। १६। ४९। ७। ५९। ८। ५६। २९। ४०।

एवं अबजदघातरूपवर्गराशेर्यावत्तावतश्च योगः अजवर्गतुल्यः अजबदघातेन[12] समानो ऽस्ति। अतो ऽयं अजवर्गतुल्यो जातः ॥ तस्मात् अजवर्गराशावेको वर्गराशिरेभिर्यावत्तावद्भिरधिको जातः। याव १ या ६। १६। ४९। ७। ५९। ८। ५६। २९। ४०।

Thus here arc AB [and] arc JD are equal to each other. And in the same way, arcs AJ [and] BD are equal. Therefore the product of AB [and] JD will be a square. Here, *BJ* [being] unknown, that is considered the amount of the *yāvattāvat*. This is multiplied by AD; result, *yā* 6; 16, 49, 7, 59, 8, 56, 29, 40.

In this way the sum of the *yāvattāvat* [times its coefficient] and the amount of the square of a number [equal to] the product of AB [and] JD is equal to the square of AJ [and thus] equal to the product of AJ [and] BD. Hence this resulting [expression is] equal to the square of AJ. Therefore in [place of] a quantity [equal to] the square of AJ, one square quantity increased by these [many] *yāvattāvats* [is] produced: *yāva* 1, *yā* 6; 16, 49, 7, 59, 8, 56, 29, 40.

[11] स्थः J_2
[12] अजबदथातेन J_2

प्रकारान्तरेण कजवर्गः साध्यते ॥ तत्र कजचापं भार्धांशेभ्यः शोध्यं शेषस्य पूर्णज्याया व्यासस्य चान्तरं कार्यम् ॥ तेन गुणितं षष्टितुल्यव्यासार्धं कबवर्गतुल्यं भविष्यति ॥ यतः कजचापं कबचापाद्विगुणमस्ति[6] ॥ तस्माद्यदि कबवर्गः याव १ षष्टि ६० तुल्यव्यासार्धेन यदा भाज्यते तदा यावद्वर्गस्य षष्ट्यंशो लभ्यते। तच्च कजचापोनभार्धांशपूर्णज्याव्यासान्तरस्य प्रमाणमस्ति ॥ पुनरिदमन्तरं संपूर्णव्यासे १२० शोधितं याव $\frac{\dot{\text{१}}}{\text{६०}}$[7] रू १२०। इदं कजचापोनभार्धंशानां पूर्णज्या भवति ॥ अस्या वर्गः याव०व० $\frac{\text{१}}{\text{३६००}}$ याव $\frac{\dot{\text{२४०}}}{\text{६०}}$[8] रू १४४००।

By another method the square of KJ is determined. There arc KJ is to be subtracted from the degrees of half a circle, [and] the difference of the chord of the remainder and the diameter is to be made. The half-diameter equal to sixty multiplied by that will be equal to the square of KB; since the arc KJ is two times the arc KB. Therefore if the square of KB is *yāva* 1, when it is divided by the half-diameter equal to sixty (60), then a sixtieth part of the square of the *yāvattāvat* is obtained. And that is the amount of the difference of the diameter and the chord of the degrees of a half-circle diminished by arc KJ. Furthermore, this difference [is] subtracted from the full diameter 120: *yāva* $-\frac{1}{60}$, *rū* 120. This is the chord of the degrees of half a circle diminished by arc KJ. Its square: *yāvava* $\frac{1}{3600}$, *yāva* $-\frac{240}{60}$, *rū* 14400.

अथ प्रकारान्तरेणाजवर्गः साध्यते ॥ तत्र अजचापं भार्धांशेभ्यः शोध्यं शेषस्य पूर्णज्याया व्यासस्य चान्तरं कार्यम्। तेन घ्नं[13] षष्टितुल्यव्यासार्धं अबवर्गतुल्यं भविष्यति ॥ यतः अजचापं अबचापाद्विगुणमस्ति ॥ तस्माद्यदि अबवर्गः याव १ षष्टि ६० तुल्यव्यासार्धेन भाज्यते तदा यावद्वर्गस्य षष्ट्यंशो लभ्यते। तच्च अजचोपोनभार्धानां या पूर्णज्या तदूनो यो व्यासस्तस्य प्रमाणमस्ति ॥ पुनरिदमन्तरं संपूर्णव्यासे १२० शोधितं याव $\frac{\dot{\text{१}}}{\text{६०}}$ रू १२०। इदं अजचापोनभार्धांशानां पूर्णज्या जाता ॥ अस्या वर्गः यावव $\frac{\text{१}}{\text{३६००}}$ याव $\frac{\dot{\text{२४०}}}{\text{६०}}$ रू १४४००।

Now by another method the square of AJ is determined. There arc AJ is to be subtracted from the degrees of half a circle, [and] the difference of the chord of the remainder and the diameter is to be made. The half-diameter equal to sixty multiplied by that will be equal to the square of AB; since the arc AJ is two times the arc AB. Therefore if the square of AB is *yāva* 1 [and is] divided by the half-diameter equal to sixty (60), then a sixtieth part of the square of the *yāvattāvat* is obtained. And that is: whatever the chord of the degrees of a half-circle diminished by arc AJ: whatever the diameter diminished by that [chord]: the amount of that. Furthermore, this difference is subtracted from the full diameter 120: *yāva* $-\frac{1}{60}$, *rū* 120. This is the resulting chord of the degrees of half a circle diminished by arc AJ. Its square: *yāvava* $\frac{1}{3600}$, *yāva* $-\frac{240}{60}$, *rū* 14400.

6 ०चापाद्वि० J_1
7 $\frac{\text{१}}{\text{६०}}$ J_1
8 $\frac{\text{२४०}}{\text{६०}}$ J_1

13 घ्न J_2

अथ च यदीष्टचापपूर्णज्यावर्गो व्यासवर्गाच्छोध्यते[9] तत्र यच्छेशं[10] तच्चापोनभार्धांशानां पूर्णज्यावर्गो भवति ॥ यतो व्यासेनेष्टचापपूर्णज्ययेष्टचापोनभा| [p. 1062 J_S] र्धांशानां पूर्णज्यया चैकं समकोणत्रिभुजं भवति ॥ यतो वृत्तार्धे व्यासप्रान्तादुत्पन्नत्रिभुजस्य पालिकोणः समकोणो भवतीति रेखागणितस्य तृतीयाध्याय उपपन्नमस्ति ॥ तत्त्रिभुजे समकोणसन्मुखभुजो व्यासो ऽस्ति ॥ स च कर्णरूपः ॥

पुनः कर्णवर्गो भुजकोट्योर्वर्गयोगेन[11] समो| [f. 3r J_1] भवतीति रेखागणितस्य प्रधमाध्याय उपपन्नम् ॥ अतो ऽत्र व्यासवर्गः १४४०० कर्णवर्गरूपः ॥ अस्मिन्निष्टचापोनभार्धांशानां

पूर्णज्यावर्गः यावव $\genfrac{}{}{0pt}{}{\text{१}}{\text{३६c०}}$ याव $\genfrac{}{}{0pt}{}{\dot{\text{२४०}}}{\text{६०}}$ [12] रू १४४०० शोधितः।

शेषं कजपूर्णज्यावर्गो ऽवशिष्टः[13] यावव $\genfrac{}{}{0pt}{}{\dot{\text{१}}}{\text{३६००}}$ याव $\genfrac{}{}{0pt}{}{\text{२४०}}{\text{६०}}$ अयं द्वितीयप्रकारेण कजवर्गः सिद्धः ॥

And now, if the square of the chord of the desired arc is subtracted from the square of the diameter, then whatever is the remainder is the square of the chord of the degrees of a half-circle diminished by that arc. Since one right-angled triangle is [formed] by the diameter and the chord of the desired arc and the chord of the degrees of a half-circle diminished by the desired arc (because "the circumference-angle of a triangle produced from the end[s] of the diameter in a half-circle is a right angle" is demonstrated in the third chapter of *Rekhāgaṇita*), the side opposite the right angle in that triangle is the diameter. And it has the form of a hypotenuse.

Furthermore, "the square of the hypotenuse is equal to the sum of the squares of the arm and the upright" is demonstrated in the first chapter of *Rekhāgaṇita*. Hence, here the square of the diameter 14400 has the form of the square of a hypotenuse. The square of the chord of the degrees of a half-circle diminished by the desired arc—*yāvava* $\frac{1}{3600}$, *yāva* $-\frac{240}{60}$, *rū* 14400—is subtracted from it. The remainder is the remaining square of the chord of KJ: *yāvava* $-\frac{1}{3600}$, *yāva* $\frac{240}{60}$. This square of KJ is established by the second method.

9 व्यासवर्ग्गाछोध्यते J_1

10 यछेषं J_1

11 ०कोट्योवर्ग्गायोगेन J_1

12 $\genfrac{}{}{0pt}{}{\text{२४०}}{\text{६०}}$ J_1

13 ०वर्गोऽशिष्टः J_1

अथ च यदीष्टचापपूर्णज्यावर्गो व्यासवर्गाच्छोध्यते तत्र शेषं| [f. 1v J_2] तच्चापोनभार्धांशपूर्णज्यावर्गो भवति ॥ यतो व्यासेनेष्टचापे पूर्णज्ययेष्टचापोनभार्धांशपूर्णज्यया चैकं समकोणत्रिभुजं भवति। यतो वृत्तार्धे व्यासप्रान्तादुत्पन्नत्रिभुजस्य पालिकोणः समकोणो भवतीत्युक्लीदसस्य तृतीयाध्याय उपपन्नमस्ति ॥ तत्त्रिभुजे समकोणसन्मुखभुजो व्यासो ऽस्ति ॥ स च कर्णरूपः ॥

पुनः कर्णवर्गो भुजकोट्योर्वर्गयोगेन समो भवतीत्युक्लीदसस्य प्रथमाध्याय[14] उपपन्नम् ॥ अतो ऽत्र व्यासवर्गः १४४०० कर्णवर्गरूपः ॥ अस्मिन्निष्टचापोनभार्धांशानां पूर्णज्यावर्गः

यावव $\genfrac{}{}{0pt}{}{\text{१}}{\text{३६००}}$ याव $\genfrac{}{}{0pt}{}{\dot{\text{२४०}}}{\text{६०}}$ रू १४४०० शोधितः। शेषं अजपूर्णज्यावर्गो ऽवशिष्टः यावव $\genfrac{}{}{0pt}{}{\dot{\text{१}}}{\text{३६००}}$ याव[15] $\genfrac{}{}{0pt}{}{\text{२४०}}{\text{६०}}$ अयं द्वितीयप्रकारेण ऽजवर्गः सिद्धः ॥

And now, if the square of the chord of the desired arc is subtracted from the square of the diameter, then the remainder is the square of the chord of the degrees of a half-circle diminished by that arc. Since one right-angled triangle is [formed] by the diameter and the chord in the desired arc and the chord of the degrees of a half-circle diminished by the desired arc (because "the circumference-angle of a triangle produced from the end[s] of the diameter in a half-circle is a right angle" is demonstrated in the third chapter of Uklīdasa (Euclid)), the side opposite the right angle in that triangle is the diameter. And it has the form of a hypotenuse.

Furthermore, "the square of the hypotenuse is equal to the sum of the squares of the arm and the upright" is demonstrated in the first chapter of Uklīdasa. Hence, here the square of the diameter 14400 has the form of the square of a hypotenuse. The square of the chord of the degrees of a half-circle diminished by the desired arc—*yāvava* $\frac{1}{3600}$, *yāva* $-\frac{240}{60}$, *rū* 14400—is subtracted from it. The remainder is the remaining square of the chord of AJ: *yāvava* $-\frac{1}{3600}$, *yāva* $\frac{240}{60}$. This square of AJ is established by the second method.

14 प्रथमध्याये J_2

15 या J_2

अथ प्रथमप्रकारागतकजवर्गः याव९ या ६। १६। ४९। ७। ५९। ८। ५६। २९। ४०[14]। एतौ समाविति समशोधनार्थं न्यासः

यावव १̇/३६००	याव २४०/६०	या ०[15]
यावव ०	याव १	या ६। १६। ४९। ७। ५९। ८। ५६। २९। ४०[16]॥

अथ बीजगणिते समीकरणसम्प्रदायस्त्वनया रीत्यास्ति ॥ सा यथा ॥ यदि समययोः पक्षयोर्मध्ये एको राशि ऋणश्चेत्तद्राशितुल्यं पक्षद्वये योज्यं तदापि पक्षद्वयं सममेव भवति ॥

तस्मादत्र प्रथमपक्षे यावव १̇/३६०० इदं पक्षद्वये[17] योज्यम् ॥ तत्र प्रथमपक्षे धनर्णयोस्तुल्यत्वाद्यावद्वर्गवर्गस्य नाशः। अवशिष्टं याव २४०/६० । अयं छेदभक्तः याव ४। |[p. 1063 J_S] अयं प्रथमपक्षः ॥

Now the square of *KJ* derived by the first method: *yāva* 1, *yā* 6; 16, 49, 7, 59, 8, 56, 29, 40. These two are equal; thus for the sake of subtraction of equals, the layout:

yāvava $-\frac{1}{3600}$	*yāva* $\frac{240}{60}$	*yā* 0
yāvava 0	*yāva* 1	*yā* 6; 16, 49, 7, 59, 8, 56, 29, 40

However, in algebra the established usage for equations is by means of this practice. That is as [follows]: When [there is] one quantity on [both] equal sides [of an equation], if it is negative, [an amount] equal to that quantity is to be added to both sides, and then the two sides are still equal.

Therefore here this *yāvava* $-\frac{1}{3600}$ on the first side is to be added to both sides. Then in the first side, because of equality of positive and negative, there is elimination of the square of the square of the *yāvat*. Remaining: *yāva* $\frac{240}{60}$. This is divided by the divisor: *yāva* 4. This is the first side.

14 ०८। ५७। ४९। ४० J_1

15 ६। ४९। ७। ५९। ८। ५७। ४९। ४० struck out J_1

16 ०८। ५७। ४९। ४० J_1

17 पक्षद्वयं J_1

अथ प्रथमप्रकारागत अजवर्गः याव १ या ६। १६। ४९। ७। ५९। ८। ५६। २९। ४०। एतौ समाविति समशोधनार्थं न्यासः

यावव १̇/३६००	याव २४०/६०	या ०
यावव ०	याव १	या ६। १६। ४९। ७। ५९। ८। ५६। २९। ४०

अथ यावनीयबीजगणिते समीकरणसम्प्रदायस्त्वनया रीत्यास्ति। सा यथा ॥ यदि समयोः पक्षयोर्मध्यो एको राशि[16] ऋणश्चेत्तद्राशितुल्यं पक्षद्वये योज्यं[17] तदापि पक्षद्वयं सममेव भवति ॥

तस्मादत्र प्रथमपक्षे यावव १̇/३६०० इदं पक्षद्वये योज्यम्। तत्र प्रथमपक्षे धनर्णयोस्तुल्यत्वाद्यावद्वर्गवर्गस्य नाशः। अवशिष्टः याव २४०/६० अयं छेदभक्तः याव ४। अयं प्रथमपक्षः।

Now the square of *AJ* derived by the first method: *yāva* 1, *yā* 6; 16, 49, 7, 59, 8, 56, 29, 40. These two are equal; thus for the sake of subtraction of equals, the layout:

yāvava $-\frac{1}{3600}$	*yāva* $\frac{240}{60}$	*yā* 0
yāvava 0	*yāva* 1	*yā* 6; 16, 49, 7, 59, 8, 56, 29, 40

However, in Yāvanī algebra the established usage for equations is by means of this practice. That is as [follows]: When [there is] one quantity on [both] equal sides [of an equation], if it is negative, [an amount] equal to that quantity is to be added to both sides, and then the two sides are still equal.

Therefore here this *yāvava* $-\frac{1}{3600}$ on the first side is to be added to both sides. Then in the first side, because of equality of positive and negative, there is elimination of the square of the square of the *yāvat*. Remaining: *yāva* $\frac{240}{60}$. This is divided by the divisor: *yāva* 4. This is the first side.

16 राशिः J_2

17 पक्षद्वयो जोज्यं J_2

पुनर्द्वितीयपक्षे यावव १/३६०० योजितः। यावव १/३६०० याव १ या ६। १६। ४९। ७। ५९। ८। ५६। २९। ४०[18]। एतौ पक्षौ पुनरपि समानौ जातौ

याव ४ या ०

यावव १/३६०० याव १ या ६। १६। ४९। ७। ५९। ८। ५६। २९। ४०[19]॥

समपक्षयोर्मध्ये यौ राश्येकजाती भवतः। तत्र लघुराशिर्महद्राशौ शोध्यते तदापि पक्षौ समावेवावशिष्टौ भवतः ॥ तस्मादत्र या वर्गराशिरेकजाती[20] यो ऽस्त्यतो लघुराशिः याव १ महद्राशौ याव ४ शोधितः। शेषं प्रथमपक्षः याव ३ द्वितीयपक्षे च यावव १/३६०० या ६। १६। ४९। ७। ५९। ८। ५६। २९। ४०[21]। एतावपि समौ ॥

Again in the second side *yāvava* $\frac{1}{3600}$ is added: *yāvava* $\frac{1}{3600}$, *yāva* 1, *yā* 6; 16, 49, 7, 59, 8, 56, 29, 40. Those two sides are again produced equal:

yāva 4 *yā* 0

yāvava $\frac{1}{3600}$ *yāva* 1 *yā* 6; 16, 49, 7, 59, 8, 56, 29, 40

In the two equal sides there are two quantities which are of the same type. There the lesser quantity is subtracted from the greater quantity, and then the two remaining sides are still equal. Therefore here whatever is a square quantity is what [we designate] the same type; therefore the lesser quantity *yāva* 1 is subtracted from the greater quantity *yāva* 4. The remainder is the first side *yāva* 3, and on the second side *yāvava* $\frac{1}{3600}$, *yā* 6; 16, 49, 7, 59, 8, 56, 29, 40. And these two are equal.

पुनर्द्वितीयपक्षे यावव १/३६०० योजितः। यावव १/३६०० याव १ या ६। १६। ४९। ७। ५९। ८। ५६। २९। ४०। एतौ पक्षौ पुनरपि समौ

याव ४ या ०

यावव १/३६०० याव १ या ६। १६। ४९। ७। ५९[18]। ८। ५६। २९। ४०

अथ यावनीयसम्प्रदाये समपक्षयोर्मध्ये यौ राश्येकजाती भवतः ॥ तत्र लघुराशिर्महद्राशौ शोध्यते तदापि पक्षौ समावेवावशिष्टौ भवतः ॥ तस्मादत्र यावर्गराशिरेकजाती[19] यो ऽस्त्यतो लघुराशिः याव १ महद्राशौ याव ४ शोधितः। शेषं प्रथमपक्षः याव ३ द्वितीयपक्षे च यावव १/३६०० या ६। १६। ४९। ७। ५९। ८। ५६। २९। ४०। एतावपि समौ ॥

Again in the second side *yāvava* $\frac{1}{3600}$ is added: *yāvava* $\frac{1}{3600}$, *yāva* 1, *yā* 6; 16, 49, 7, 59, 8, 56, 29, 40. Those two sides are again equal:

yāva 4 *yā* 0

yāvava $\frac{1}{3600}$ *yāva* 1 *yā* 6; 16, 49, 7, 59, 8, 56, 29, 40

Now in Yāvanī established usage, in the two equal sides there are two quantities which are of the same type. There the lesser quantity is subtracted from the greater quantity, and then the two remaining sides are still equal. Therefore here whatever is a square quantity is what [we designate] the same type; therefore the lesser quantity *yāva* 1 is subtracted from the greater quantity *yāva* 4. The remainder is the first side *yāva* 3, and on the second side *yāvava* $\frac{1}{3600}$, *yā* 6; 16, 49, 7, 59, 8, 56, 29, 40. And these two are equal.

---

18 ०८। ५७। ४९। ४० J_1
19 ०८। ५७। ४९। ४० J_1
20 यो वर्गः राशिः एकजाती J_1
21 ०८। ५७। ४९। ४० J_1

---

18 ५९ missing in J_2
19 यो राशि० J_2

पुनः पक्षद्वयमध्ये यद्येकः खण्डितो ऽस्ति द्वितीयपक्षः| [f. 3v J_1] संपूर्णश्चेत्तदा खण्डितराशिं प्रपूर्य तावद्गुणं द्वितीयपक्षमपि कुर्वन्ति ॥ अत्र सच्छेदराशिः खण्डित उच्यते छेदरहितः संपुर्ण उच्यते ॥ एवं कृते सति पक्षयोः समछेदत्वं[22] विधाय छेदापगम एवोपपद्यते ॥

तस्मादत्र द्वितीयपक्षे खण्डितराशिः यावव $\frac{१}{३६००}$। अयमर्थः ॥ अत्र यावद्वर्गवर्गराशिरस्ति[23]। तस्मादयं हरगुणितः[24] क्रियते। यावत्तावद्वर्गवर्गराशिरेको[25] भवतीति द्वितीयपक्षो ऽप्येतावता गुणनीयः ॥ एवं कृते प्रथमपक्षे याव १०८०० द्वितीयपक्षे यावव १। अथ प्रथमपक्षो[26] वारद्वयं षष्ट्योर्ध्वीकृतः याव ३। अथ द्वितीयपक्षस्य यावद्राशिस्तेनैव छेदेन ३६०० संगुण्य वारद्वयं षष्ट्योर्ध्वीकृतः या ६। १६। ४९। ७। ५९। ८। ५६। २९। ४०।

Moreover, if [there] is one broken [quantity] among the two sides, if the second side is complete, then having completed [the side with] the broken quantity, they make the second side also multiplied by that much. Here a quantity with divisor is called "broken" [and one] without divisor is called "complete". When it is done thus [and one has] established equality of divisors in the two sides, the elimination of divisors is achieved.

Therefore here on the second side the broken quantity is *yāvava* $\frac{1}{3600}$. This is the meaning. Here [there is] a coefficient of the square of the square of the *yāvattāvat*; therefore this is multiplied by the denominator. The coefficient of the square of the square of the *yāvattāvat* is one, so the second side is also to be multiplied by so much. When done thus, on the first side [there is] *yāva* 10800 [and] on the second side *yāvava* 1. Now the first side is sixty-raised two places: *yāva* 3. Now the coefficient of the *yāvattāvat* of the second side is multiplied by that very divisor 3600, [i.e.,] sixty-raised two places: *yā* 6, 16, 49; 7, 59, 8, 56, 29, 40.

22 समछेदत्व J_1
23 यावद्वर्गवर्गवर्गराशिरस्ति J_1
24 हरं गुणितः J_1
25 तावद्यावद्वर्गराशि∘ J_1
26 प्रमपक्षः J_1

पुनः[20] पक्षद्वयमध्ये यद्येकः पक्षः खण्डितो द्वितीयपक्षः संपूर्णश्चेत्तदा खण्डितराशिं प्रपूर्य[21] तावद्गुणं द्वितीयपक्षमपि कुर्वन्ति ॥ अत्र सच्छेदराशिः खण्डितशब्देनोच्यते। छेदरहितः संपूर्णोच्यते ॥ एवं कृते सति पक्षयोः समछेदत्वं विधाय छेदापगम एवोपपद्यते ॥

तस्मादत्र द्वितीयप|[f. 2r J_2] क्षे खण्डितराशिः यावव $\frac{१}{३६००}$। अयमर्थः। अत्र यावद्वर्गवर्गराशेः[22] षड्शताधिकसहस्रत्रयमितोंऽशो यावद्वर्गवर्गराशिरस्ति। तस्मादयं हरगुणितः[23] क्रियते। यावत्तावद्वर्गवर्गराशिरेको[24] भवतीति[25] द्वितीयपक्षो ऽप्येतावता गुणनीयः[26] ॥ एवं कृते प्रथमपक्षे[27] याव १०८०० द्वितीयपक्षे यावव १। अथ प्रथमपक्षो वारद्वयं षष्ट्योर्ध्वीकृतः[28] याव ३। अथ द्वितीयपक्षस्थयावद्राशिस्तेनैव छेदेन ३६०० संगुण्य वारद्वयं षष्ट्योर्ध्वीकृतः[29] या ६। १६। ४९। ७। ५९। ८। ५६। २९। ४०।

Moreover, if [there is] one broken side among the two sides, if the second side is complete, then having completed [the side with] the broken quantity, they make the second side also multiplied by that much. Here a quantity with divisor is called by the term "broken"; [one] without divisor is called "complete". When it is done thus [and one has] established equality of divisors in the two sides, the elimination of divisors is achieved.

Therefore here on the second side the broken quantity is *yāvava* $\frac{1}{3600}$. This is the meaning. Here the coefficient of the square of the square of the *yāvattāvat* is a part equal to three thousand plus six hundred of the amount of the square of the square of the *yāvattāvat*; therefore this is multiplied by the denominator. The coefficient of the square of the square of the *yāvattāvat* is one, so the second side is also to be multiplied by so much. When done thus, on the first side [there is] *yāva* 10800 [and] on the second side *yāvava* 1. Now the first side is sixty-raised two places: *yāva* 3. Now the coefficient of the *yāvattāvat* in the second side is multiplied by that very divisor 3600, [i.e.,] sixty-raised two places: *yā* 6, 16, 49; 7, 59, 8, 56, 29, 40.

20 पुन J_2
21 प्रपूर्य्यं J_2
22 ∘राशे J_2
23 हरगुणितं J_2
24 ताव १ द्यावद्वर्गवर्गराशिरेको J_2
25 ति inserted from left margin J_2
26 द्वितीयपक्षमप्येतावता गुणनीयं J_2
27 प्रथमक्षे J_2
28 षष्टोर्ध्वीकृतः J_2
29 षष्ट्यो उर्ध्वी कृतः J_2

एवं जातौ पक्षौ समौ।

याव ३ द्विपरि|[p. 1064 J_S] वर्ताः
यावव १ या ६ । १६ । ४९ । ७ । ५९ । ८ । ५६[27] । २९ । ४०

एतौ यावत्तावतापवर्तितौ ॥

या ३ द्विपरिवर्ताः
याघ १ रू ६ । १६ । ४९ । ७ । ५९ । ८ । ५६ । २९[28] । ४०

अत्र यावद्घनः सरूपराशिश्चेद्यावत्तावत्त्रयेण ३[29] भाज्यते तदा यावत्तावत्प्रमाणं लभ्यते। अस्य भागग्रहणरीतिः पूर्वाचार्यैरेतावत्कालपर्यन्तं न लब्धा ॥

In this way the two resulting equal sides are:

yāva 3 twice-shifted
yāvava 1, *yā* 6, 16, 49; 7, 59, 8, 56, 29, 40

These two are reduced by a *yāvattāvat*:

yā 3 twice-shifted
yāgha 1, *rū* 6, 16, 49; 7, 59, 8, 56, 29, 40

Here if the cube of *yāvattāvats* [and] the quantity with units is divided by the triple [coefficient] of the *yāvattāvat*, 3, then the amount of the *yāvattāvat* is obtained.

His technique of obtaining the quotient [has] not [been] arrived at by the previous teachers up to now.

[27] ०७ । ५८ । ५६ J_1
[28] ०८ । ५७ । २९ J_1
[29] या ३ J_1

एवं जातौ पक्षौ समौ।

याव ३ द्विपरिवर्ताः
यावव १ या ६ । १६ । ४९ । ७ । ५९ । ८ । ५६ । २९ । ४० द्विपरिवर्ताः

एतौ यावत्तावतापवर्तितौ[30]।

या ३ द्विपरिव०
याघ १ रू ६ । १६ । ४९ । ७ । ५९ । ८ । ५६ । २९ । ४० द्विपरिव०

अथात्र यावद्घनेन[31] सरूपस्थराशिना चेद्यावत्तावत्त्रयं[32] ३ भाज्यते तदा यावत्तावत्प्रमाणं लभ्यते। अस्य भागग्रहणरीतिः पूर्वाचार्यैरेतावत्कालपर्यन्तं न लब्धा।

In this way the two resulting equal sides are:

yāva 3 twice-shifted
yāvava 1, *yā* 6, 16, 49; 7, 59, 8, 56, 29, 40 twice-shifted

These two are reduced by a *yāvattāvat*:

yā 3 twice-shifted
yāgha 1, *rū* 6, 16, 49; 7, 59, 8, 56, 29, 40 twice-shifted

Now here, if the triple [coefficient] of the *yāvattāvat*, 3, is divided by the cube of the *yāvattāvat* together with the quantity in units [? division reversed?], then the amount of the *yāvattāvat* is obtained.

His technique of obtaining the quotient [has] not [been] arrived at by the previous teachers up to now.

[30] ०तापवत्ति तौ J_2
[31] यावद्द्वयेन J_2
[32] चेद्यावत्तावत्रयं J_2

अत्र यमशैदेन[30] रीतिः प्रदर्शिता ॥ सा यथा। रूपस्य प्रथमाङ्को यावत्तावता भाज्यः। लब्धिरेकान्ते[31] स्थाप्या ॥ पुनर्लब्धिघनं शेषाङ्के योज्यं[32] पुनरत्र द्वितीयाङ्को यावत्तावता भाज्यः। लब्धिः पूर्वलब्धेरधः स्थाप्या ॥ पुनर्लब्धिद्वययोगस्य घनः कार्यः ॥ एवं तत्र प्रथमलब्धिघनः शोध्यः। शेषं द्वितीयलब्धिशेषे योज्यम् ॥

पुनस्तृतीयाङ्को यावत्तावता भाज्यः। इयं लब्धिः पूर्वलब्धिद्वयाधः स्थाप्या ॥ पुनरस्य लब्धित्रययोगस्य घनः कार्यः। तत्र लब्धिद्वययोगस्य घनः शोध्यः [f. 4r J_1] |। शेषं तृतीयलब्धिशेषे योज्यम् ॥ पुनस्तत्र चतुर्थाङ्को यावत्तवता भाज्यः ॥ एवमेवेष्टभागपर्यन्तं विधिः कार्यः ॥

एवं यमशैदेन यावत्तावत्प्रमाणं निष्काशितं[33] २ । ५ । ३९ । २६ । २२ । २९ । २८। ३२ । ५२ । ३३ । इयमंशद्वयस्य पूर्णज्यास्ति। अस्यार्धमेकांशज्या जाता ॥ १ । २ । ४९ । ४३ । ११ । १४ । ४४ । १६ । २६ । १७।

Here the technique is expounded by Yamaśaida (Jamshīd al-Kāshī). It is as [follows]:

The first digit of the units [term] is to be divided by the *yāvattāvat* [coefficient]. The quotient is to be set at one end. Moreover, the cube of the quotient is to be added to the remainder-digit; here again, the second digit is to be divided by the *yāvattāvat* [coefficient]. The quotient is to be set below the previous quotient. Also the cube of the sum of the two quotients is to be made. In this way the cube of the first quotient is to be subtracted from that. The remainder is to be added to the remainder of the second quotient.

Once again, the third digit is to be divided by the *yāvattāvat* [coefficient]. This quotient is to be set below the previous quotient. Again, the cube of that sum of the three quotients is to be made. The cube of the sum of the two quotients is to be subtracted from that. The remainder is to be added to the remainder of the third quotient. There again, the fourth digit is to be divided by the *yāvattāvat* [coefficient]. In just this way the rule is to be performed up to the desired place.

In this way the amount of the *yāvattāvat* is extracted by Yamaśaida: 2; 5, 39, 26, 22, 29, 28, 32, 52, 33. This is the chord of two degrees. Half of it is the resulting sine of one degree: 1; 2, 49, 43, 11, 14, 44, 16, 26, 17.

[30] यमशदेन J_1
[31] लब्धिरेकांत J_1
[32] योज्य J_1
[33] निष्काशिलं J_1

अत्र यमशैदेन रीतिः प्रदर्शिता ॥ सा यथा ॥ रूपस्य प्रथमाङ्को यावत्तावता भाज्यः। लब्धिरेकान्ते स्थाप्या ॥ पुनर्लाब्धिघनं शेषाङ्के योज्यं पुनरत्र द्वितीयाङ्को यावत्तावता भाज्यः। लब्धिः पूर्वलब्धेरधः स्थाप्या[33] ॥ पुनर्लब्धिद्वययोगस्य घनः कार्यः। एवं तत्र प्रथमलब्धिघनः शोध्यः ॥ शेषं द्वितीयलब्धिशेषे योज्यम्[34] ॥

पुनस्तृतीयाङ्को यावत्तावता भाज्यः ॥ इयं लब्धिः पूर्वलब्धिद्वयाधः स्थाप्या[35] ॥ पुनरस्य लब्धित्रययोगस्य घनः कार्यः। तत्र लब्धिद्वययोगस्य घनः शोध्यः ॥ शेषं तृतीयलब्धिशेषे योज्यम् ॥ पुनस्तत्र चतुर्थाङ्को यावत्तावता भाज्यः ॥

एवमेवेष्टभागलब्धिपर्यन्तं विधिः कार्यः ॥ एवं यमशैदेन[36] यावत्तावत्प्रमाणं निष्काशितं[37] २ । ५ । ३९ । २६ । २२ । २९ । २८ । ३२ । ५२ । ३३। इयमंशद्वयस्य[38] पूर्णज्यास्ति ॥ अस्यार्धमेकांशज्या जाता १ । २ । ४९ । ४३ । ११ । १४ । ४४ । १६ । २६ । १७।

Here the technique is expounded by Yamaśaida (Jamshīd al-Kāshī). It is as [follows]:

The first digit of the units [term] is to be divided by the *yāvattāvat* [coefficient]. The quotient is to be set at one end. Moreover, the cube of the quotient is to be added to the remainder-digit; here again, the second digit is to be divided by the *yāvattāvat* [coefficient]. The quotient is to be set below the previous quotient. Also the cube of the sum of the two quotients is to be made. In this way the cube of the first quotient is to be subtracted from that. The remainder is to be added to the remainder of the second quotient.

Once again, the third digit is to be divided by the *yāvattāvat* [coefficient]. This quotient is to be set below the previous quotient. Again, the cube of that sum of the three quotients is to be made. The cube of the sum of the two quotients is to be subtracted from that. The remainder is to be added to the remainder of the third quotient. There again, the fourth digit is to be divided by the *yāvattāvat* [coefficient]. In just this way the rule is to be performed up to the quotient at the desired place.

In this way the amount of the *yāvattāvat* is extracted by Yamaśaida: 2; 5, 39, 26, 22, 29, 28, 32, 52, 33. This is the chord of two degrees. Half of it is the resulting sine of one degree: 1; 2, 49, 43, 11, 14, 44, 16, 26, 17.

[33] लब्धि पूर्वलध्वेरधस्थाप्यः J_2
[34] ज्योज्यं J_2
[35] पूर्वलब्धिद्वयाधस्थाप्यः J_2
[36] यमशदैन J_2
[37] यावत्तावत्प्रमाणो निष्काशितः J_2
[38] इयमंश० J_2

अस्य भागहरणोपपत्तिः ॥ तत्र पक्षद्वयमध्य एकपक्षे यावत्तावदस्ति ॥ द्वितीयपक्षे चावत्तावद्घनो रूपराशिश्च ॥ एवं तत्र यावत्तावद्ज्ञानं चेत्तदा यावत्तावद्घनं कृत्वा रूपराशौ प्रक्षिप्य यावत्तावता भाज्यते। तत्र लब्धिर्यावत्तावन्मानं स्यात् ॥

अत्र तु यावत्तावद्ज्ञानं नास्त्यतो रूपराशिरेव यावत्तावता भाज्यः ॥ यल्लब्धं तत्सरूपचावद्घनस्य |[p. 1065 J_S] कोऽप्यंशो लब्धः ॥ स चैकान्ते धृतः ॥ पुनरस्य घनं कृत्वा शेषे योज्यम्[34]। पुनरत्र यावत्तावता द्वितीयाङ्को भाजितः। एवं यल्लब्धं तत्पूर्वलब्धेरधः स्थापितम् ॥ एवं लब्धं यावत्तावद्घनस्यासन्नो भागो लब्धः ॥

अत्रासन्नता शेषावयवसत्वान्निरवयवत्व एव सूक्ष्मा लब्धिः ॥ अथ लब्धिघने पूर्वघनं संशोध्य यतस्तदधिकं जातमेवमिष्टभागपर्यन्तं मुहुर्वारं वारं कार्यम् ॥

The demonstration of its division process. There among the two sides, on one side is the *yāvattāvat*. On the second side are the cube of the *yāvattāvat* and the quantity in units. In this way if the *yāvattāvat* there is known, then [when one has] made the cube of the *yāvattāvat* and added it to the quantity in units, [the result] is divided by the *yāvattāvat* [coefficient]. Then the quotient is the amount of the *yāvattāvat*.

But here the *yāvattāvat* is not known; hence the quantity in units itself is to be divided by the *yāvattāvat* [coefficient]. Whatever is the quotient, that is taken [as] some part of the cube of the *yāvattāvat* increased by the units. And it is kept at one end. Furthermore, [when one has] made its cube, [that] is to be added to the remainder. Here again, the second digit is divided by the *yāvattāvat* [coefficient]. In this way, whatever is the quotient is set below the quotient previous to that. The quotient [computed] in this way is taken [as] the approximate part of the cube of the *yāvattāvat*.

Here [there is] approximate-ness; [but] when there is immutability [of the quotient], after the state of having [different] components [in the] remainder, the quotient is indeed accurate. Now [when one has] subtracted the previous cube from the cube of the [current] quotient, because the resulting [cube of the quotient] is greater than that [previous cube], in this way it is to be done again repeatedly up to the desired place.

34 योज्य J_1

अस्य भागहरणोपपत्तिः ॥ तत्र पक्षद्वयमध्य एकपक्षे यावत्तावदस्ति ॥ द्वि०पक्षे यावद्घनो रूपराशिश्च ॥ एवं तत्र यावत्तावद्ज्ञानं चेत्तदा यावत्तावद्घनं कृत्वा रूपराशौ प्रक्षिप्य यावत्तावता भाज्यते। तत्र लब्धिर्यावत्तावन्मानं स्यात् ॥

अत्र तु यावत्तावद्ज्ञानं नास्त्यतो रूपराशिरेव यावत्तावता भाज्यः। यल्लब्धं तत्सरूपयावद्घनस्य कोऽप्यंशो लब्धः। स चैकान्ते धृतः ॥ पुनरस्य घनं कृत्वा| [f. 2v J_2] शेषे योजितम्[39]। पुनरत्र यावत्तावता द्वितीयाङ्को भाजितः ॥ एवं यल्लब्धं तत्तु पूर्वलब्धेरधः स्थापितम् ॥ एवं लब्धं यावत्तावद्घनस्यासन्नो भागो लब्धः।

अत्रासन्नता शेषावयवसत्वान्निरवयवत्व[40] एव सूक्ष्मा लब्धिः ॥ अथ लब्धिघने पूर्वघनं शोध्यं यतस्तदधिकं जातम् ॥ एवमिष्टभागपर्यन्तं मुहुर्मुहुः कार्यम् ॥ ॥

The demonstration of its division process. There among the two sides, on one side is the *yāvattāvat*. On the second side are the cube of the *yāvattāvat* and the quantity in units. In this way if the *yāvattāvat* there is known, then [when one has] made the cube of the *yāvattāvat* and added it to the quantity in units, [the result] is divided by the *yāvattāvat* [coefficient]. Then the quotient is the amount of the *yāvattāvat*.

But here the *yāvattāvat* is not known; hence the quantity in units itself is to be divided by the *yāvattāvat* [coefficient]. Whatever is the quotient, that is taken [as] some part of the cube of the *yāvattāvat* increased by the units. And it is kept at one end. Furthermore, [when one has] made its cube, [that] is added to the remainder. Here again, the second digit is divided by the *yāvattāvat* [coefficient]. In this way, whatever is the quotient, that itself is set below the previous quotient. The quotient [computed] in this way is taken [as] the approximate part of the cube of the *yāvattāvat*.

Here [there is] approximate-ness; [but] when there is immutability [of the quotient], after the state of having [different] components [in the] remainder, the quotient is indeed accurate. Now the previous cube is subtracted from the cube of the [current] quotient, because the resulting [cube of the quotient] is greater than that [previous cube]. In this way it is to be done again and again up to the desired place.

39 जोतितं J_2

40 शेषावयवसत्वात् निरवयवत्वे स J_2

अथास्योदाहरणम् ॥ तत्र द्वितीयपक्षे रूपाणि ६। १६। ४९। ७। ५९। ८। ५६ । २९। ४०॥ अत्र षड्द्विपरिवर्ताः ॥ षष्टेरूर्ध्वमस्ति ॥ यावत्तावत्त्रयं द्विपरिवर्ताः षष्टेरूर्ध्वमस्ति ॥

अत एकजातौ भागे गृहीते लब्धमंशाः। २ पुनरस्य घनेंऽशाः ८ वारद्वयं षष्ट्या भक्ताः ०। ०। ८। विकलात्मकं विकलासु योजितं १६। ५७। पुनः पू|[f. 4v J_1] र्वलब्धिशेषं १६ तेनैव या ३ भक्तं लब्धाः कलाः ५। लब्धिः पूर्वलब्धे २ रधः[35] स्थापिता। अं क २। ५।

पुनरत्र शिष्टं १। ५७। पुनर्लब्धिद्वयस्य घनः अं ९। २। ३२। ५। अत्र प्रथमलब्धिघनः अं ८ शोधितः शेषं अं १। २। ३२। ५ ॥ इदं शेषे १। अं[36] ५७। ७। ५९। ८। ५६। २९। ४०। अंशस्थाने योजितं १। ५८। १०। ३१। १३। पुनर्हरेण[37] या ३ अंशे ५८ भागो गृहीतः। लब्धिर्विकलात्मिका ३९। शेषं अं १। १०। ३१। १३। पुनरियं लब्धिः पूर्वलब्धेरधः स्थाप्या[38] २। ५। ३९।

Now, its example. On the second side in it are units: 6, 16, 49; 7, 59, 8, 56, 29, 40. Here the six is twice-shifted, [that] is, raised by sixty. Three times the *yāvattāvat* is twice-shifted, raised by sixty.

Hence when the divisor is understood [to be] of the same order, the quotient is degrees: 2. Further, the degrees in its cube, 8, are divided by sixty two times: 0, 0, 8. Added in the form of seconds to seconds: 16, 57. Again, the remainder from the previous quotient, 16, [is] divided by just that *yā* 3, [and] minutes are obtained: 5. The quotient is put down below [i.e., to the right of] the previous quotient 2: ° ′ 2, 5.

Here again, the remainder: 1, 57. Again, the cube of the two[-digit] quotient: 9°, 2, 32, 5. From this the cube of the first quotient, 8°, is subtracted; the remainder: 1°, 2, 32, 5. This [is added] to the remainder: 1, 57°, 7, 59, 8, 56, 29, 40, in the degrees place: 1, 58, 10, 31, 13. Again a part is taken with the divisor *yā* 3 in the degree [place]: 58. The quotient is in the form of minutes: 39; the remainder: 1°, 10, 31, 13. Again this quotient is to be set below [i.e., after] the previous quotient: 2, 5, 39.

अथास्योदाहरणम्[41] ॥ तत्र द्वितीयपक्षे रूपाणि ६। १६। ४९। ७। ५९। ८। ५६ । २९। ४०। अत्र षट् ६ द्विपरिवर्ताः। षष्टेरूर्ध्वमस्ति ॥ यावत्तावत्त्रयं द्विपरिवर्ताः षष्टेरूर्ध्वमस्ति ॥

अत एकजातौ भागे गृहीते लब्धमंशाः २। पुनरस्य घनेंऽशाः ८ शेषां १६ शेषु ४९ योजितं १६। ५७। पुनः पूर्वलब्धिशेषं १६ तेनैव या ३ भक्तं लब्धं कलाः ५। लब्धिः पूर्वलब्धेरधः स्थापिता। अं २ क ५।

पुनरत्र शिष्टं १। ५७। पुनर्लब्धिद्वयस्य घनः अं ९। २। ३२। ५। अत्र प्रथमलब्धिघनः अं ८ शोधितः शेषं अं १। २। ३२। ५[42] । इदं शेषे १। अं ५७। क ७। वि ५९। ८। ५६। २९। ४०। अंशस्थाने योजितं १। अं ५८। १०। ३१। १३। पुनर्हरेण या ३ अंशे ५८[43] भागो गृहीतः। लब्धिर्विकलात्मिका ३९। शेषं अं १। १०। ३१। १३। पुनरियं लब्धिः पूर्वलब्धेरधः स्थाप्या[44] अं २। ५। ३९।

Now, its example. On the second side in it are units: 6, 16, 49; 7, 59, 8, 56, 29, 40. Here the six (6) is twice-shifted, [that] is, raised by sixty. Three times the *yāvattāvat* is twice-shifted, raised by sixty.

Hence when the divisor is understood [to be] of the same order, the quotient is degrees: 2. Further, the degrees in its cube, 8, are added to the degrees of the remainder, 16, 49: 16, 57. Again, the remainder from the previous quotient, 16, [is] divided by just that *yā* 3, [and] the quotient is minutes: 5. The quotient is put down below [i.e., to the right of] the previous quotient: 2°, 5′.

Here again, the remainder: 1, 57. Again, the cube of the two[-digit] quotient: 9°, 2, 32, 5. From this the cube of the first quotient, 8°, is subtracted; the remainder: 1°, 2, 32, 5. This [is added] to the remainder: 1, 57°, 7′, 59″, 8, 56, 29, 40, in the degrees place: 1, 58, 10, 31, 13. Again a part is taken with the divisor *yā* 3 in the degree [place]: 58. The quotient is in the form of minutes: 39; the remainder: 1°, 10, 31, 13. Again this quotient is to be set below [i.e., after] the previous quotient: 2°, 5, 39.

35 पूर्वलब्धेः २ अधः J_1

36 अं १। J_1

37 पुनहरेण J_1

38 स्थाप्य J_1

41 अथास्योदोहरणं J_2

42 ५७ J_2

43 ५८ above line J_2

44 स्थाप्यः J_2

अस्य घनोंऽशादिः |[p. 1066 J_S] ९ । ११ । २ । ३२ । २७ । ४३ । ३९॥ [f. 5r J_1] | अत्र[39] द्वितीयघनः ९ । २ । ३२ । ५ शोधितः। शेषं कलादि ८ । ३०। २७ । २७ । ४३ । ३९। इदं शेषाङ्केषु १ । १० । ३१ । १३ योजितं १ । १९। १ । ४१ । २४ । १३ । १९। पुनरत्र कलास्थाने[40] भक्तं ल॰ २६ पूर्वलब्धेरधः स्थापिता २ । ५ । ३९ । २६।

शेषं कलादि १ । १ । ४१ । २४। १३ । १९। पुनर्लब्धैः २ । ५ । ३९ । २६ घनः ९ । ११ । ८ । १४ । ३३ । १२। २३ । २१ । ४ । ५६। तृतीयघनो ऽत्र शोधितः। विकलास्ववशिष्टं ५ । ४२ । ५ । २८ । ४४ । २१ । ४ । ५६। पुनरिदं शेषाङ्के योजितं क १ । ७ । २३ । २९ । ४२ । ३ । २१ । ४ । ५६। पुनस्तेनैव भक्ते लब्धं २२ शेषं १ । २३ । २९ । ४२ । ३ । २१ । ४ । ५६। पूर्वलब्धौ योजितं २ । ५ । ३९ । २६ । २२।

Its cube in degrees etc.: 9, 11, 2, 32, 27, 43, 39. The cube of the second [quotient] 9, 2, 32, 5 is subtracted from this. The remainder in minutes etc.: 8, 30, 27, 27, 43, 39. This is added to the digits of the remainder, 1, 10, 31, 13: [result,] 1, 19, 1, 41, 24, 13, 19. Again here [that remainder is] divided in the minutes place: quo. 26, set below the previous quotient: 2, 5, 39, 26.

The remainder in minutes etc.: 1, 1, 41, 24, 13, 19. Again, with the obtained [digits] 2, 5, 39, 26, the cube: 9, 11, 8, 14, 33, 12, 23, 21, 4, 56. The cube of the third [quotient] is subtracted from this. The remainder in seconds: 5, 42, 5, 28, 44, 21, 4, 56. Again, this is added to the number of the remainder: 1′, 7, 23, 29, 42, 3, 21, 4, 56. When divided again by that same [divisor], quotient: 22, remainder: 1, 23, 29, 42, 3, 21, 4, 56. Added to the previous quotient: 2, 5, 39, 26, 22.

[39] त्र inserted from top margin J_1

[40] कलास्थानें J_1

अस्य घनोंऽशादिः ९ । ११ । २ । ३२ । २७ । ४३ । ३९। अत्र द्वितीयघनः ९ । २ । ३२ । ५ शोधितः । शेषं कलादि ८ । ३० । २७ । २७ । ४३ । ३९। इदं शेषाङ्केषु अ १ । १० । ३१ । १३ योजितं अं १ । १९ । १ । ४१ । २४ । १३ । १९। पुनरत्र कलास्थाने भक्तं ल॰ २६ पूर्वलब्धेः स्थापिता २ । ५ । ३९ । २६।

शेषं कलादि १ । १ । ४१ । २४ । १३ । १९। पुनर्लब्धेर्धनः[45] ९ । ११ । ८ । १४ । ३३ । १२ । २३ । २१ । ४ । ५६। तृतीयधनो ऽत्र शोधितः। विकलाद्यवशिष्टं[46] वि ५ । ४२ । ५ । २८ । ४४ । २१ । ४ । ५६। पुनरिदं शेषाङ्के योजितं क १ । ७ । २३ । २९ । ४२ । ३ । २१ । ४ । ५६। पुनस्तेनैव भक्ते लब्धं २२ शेषं १ । २३ । २९ । ४२ । ३ । २१ । ४ । ५६। पूर्वलब्धियोजितं २ । ५ । ३९ । २६ । २२।

Its cube in degrees etc.: 9, 11, 2, 32, 27, 43, 39. The cube of the second [quotient] 9, 2, 32, 5 is subtracted from this. The remainder in minutes etc.: 8, 30, 27, 27, 43, 39. This is added to the digits of the remainder, 1°, 10, 31, 13: [result,] 1°, 19, 1, 41, 24, 13, 19. Again here [that remainder is] divided in the minutes place: quo. 26, set [below] the previous quotient: 2, 5, 39, 26.

The remainder in minutes etc.: 1, 1, 41, 24, 13, 19. Again, the cube of the quotient: 9, 11, 8, 14, 33, 12, 23, 21, 4, 56. The cube of the third [quotient] is subtracted from this. The remainder in seconds etc.: 5″, 42, 5, 28, 44, 21, 4, 56. Again, this is added to the number of the remainder: 1′, 7, 23, 29, 42, 3, 21, 4, 56. When divided again by that same [divisor], quotient: 22, remainder: 1, 23, 29, 42, 3, 21, 4, 56. Added to the previous quotient: 2, 5, 39, 26, 22.

[45] पुनर्लब्धे । र्धनः J_2

[46] द्य inserted from right margin J_2

अस्य घनः ९। ११। ८। १९। २२। ४१। ६। ५२। १५। १। ५६। चतुर्थघनो ऽत्र शोधितः शेषं ४। ४९। २८। ४३। ३१। १०। ५। ५६। इदं शेषे योजितं १। २८। १९। १०। ४६। ५२। १५। १। ५६। पुनर्हरभक्तः। लब्धं २९ शेषं १। १९। १०। ४६। ५२। १५। १। ५६। पुनर्लब्धिरधः स्थिता २। ५। ३९। २६। २२। २९।

अस्य घनः ९। ११। ८। १९। २९। २। ४२। १। ३९। ५०। ५२। अत्र पञ्चमघनः[41] शोधितः शेषं ६। २१। ३५। ९। २४। ४८। ५६। इदं शेषे योजितं १। २५। ३२। २२। १। ३९। ५०। ५२। पुनरयं हरया३भक्तः। लब्धं २८ शेषं १। ३२। २२। १। ३९। ५०। ५२। लब्धिः पूर्वलब्धेरधः स्थापिता २। ५। ३९। २६। २२। २९। २८।

अस्य घनः ९। ११। ८। १९। २९। ८। ५०। २७। १९। ५९। ४३। अत्र षष्टघनः शोधितः शेषं ६। ८। २५। ४०। ८। ५१। शेषे योजितं १। ३८। ३०। २७। १९। ५९। ४३। पुनर्हरभक्तं लब्ध ३२|[f. 5v J_1] शेषं २। ३०। २७। १९। ५९। ४३। लब्धिः पूर्वलब्धेरधः स्थापिता २। ५। ३९। २६। २२। २९। २८। ३२।

अस्य घनः अं ९। ११। ८। १९। २२। ४१। ६। ५२। १५। १। ५६। चतुर्थघनो ऽत्र शोधितः शेषं ४। ४९। २८। ४३। ३१। १०। ५। ५६। इदं शेषे योजितं १। २८। १९। १०। ४६। ५२। १५। १। ५६। पुनर्हरभक्तः। लब्धं २९ शेषं १। १९। १०। ४६। ५२। १५। १। ५६। पुनर्लब्धिरधः स्थिता[47] २। ५। ३९। २६। २२। २९।

अस्य घनः ९। ११। ८। १९। २९। २। ४२। १। ३९। ५०। ५२। अत्र पञ्चमघनः शोधितः शेषः ६। २१। ३५। ९। २४। ४८। ५६। अयं[48] शेषे योजितः[49] १। २५। ३२। २२। १। ३९। ५०। ५२। पुनरयं हरया३भक्तः। ल २८ शेषं १। ३२। २२। १। ३९। ५०। ५२। लब्धिः पूर्वलब्धेरधः स्थापिता[50] २। ५। ३९। २६। २२। २९। २८।

अस्य घनः ९। ११। ८। १९। २९। ८। ५०। २७। १९। ५९। ४३। अत्र षष्टघनः शोधितः शो|[f. 3r J_2] षं ६। ८। २५। ४०। ८। ५१। शेषे योजितं[51] १। ३८। ३०। २७। १९। ५९। ४३। पुनर्हरभक्तः। ल० ३२ शे २। ३०। २७। १९। ५९। ४३। लब्धिः पूर्वलब्धेरधः स्थापिता २। ५। ३९। २६। २२। २९। २८। ३२।

Its cube: 9, 11, 8, 19, 22, 41, 6, 52, 15, 1, 56. The cube of the fourth [quotient] is subtracted from this; remainder: 4, 49, 28, 43, 31, 10, 5, 56. This is added to the remainder: 1, 28, 19, 10, 46, 52, 15, 1, 56. Divided again by the divisor, quotient: 29, remainder: 1, 19, 10, 46, 52, 15, 1, 56. Again, the quotient is set below: 2, 5, 39, 26, 22, 29.

Its cube: 9, 11, 8, 19, 29, 2, 42, 1, 39, 50, 52. The cube of the fifth [quotient] is subtracted from this; remainder: 6, 21, 35, 9, 24, 48, 56. This is added to the remainder: 1, 25, 32, 22, 1, 39, 50, 52. Again, this is divided by the divisor *yā* 3. Quotient: 28, remainder: 1, 32, 22, 1, 39, 50, 52. The quotient is set below the previous quotient: 2, 5, 39, 26, 22, 29, 28.

Its cube: 9, 11, 8, 19, 29, 8, 50, 27, 19, 59, 43. The cube of the sixth [quotient] is subtracted from this; remainder: 6, 8, 25, 40, 8, 51. Added to the remainder: 1, 38, 30, 27, 19, 59, 43. Again divided by the divisor, quotient: 32, remainder: 2, 30, 27, 19, 59, 43. The quotient is set below the previous quotient: 2, 5, 39, 26, 22, 29, 28, 32.

Its cube: 9°, 11, 8, 19, 22, 41, 6, 52, 15, 1, 56. The cube of the fourth [quotient] is subtracted from this; remainder: 4, 49, 28, 43, 31, 10, 5, 56. This is added to the remainder: 1, 28, 19, 10, 46, 52, 15, 1, 56. Divided again by the divisor, quotient: 29, remainder: 1, 19, 10, 46, 52, 15, 1, 56. Again, the quotient is set below: 2, 5, 39, 26, 22, 29.

Its cube: 9, 11, 8, 19, 29, 2, 42, 1, 39, 50, 52. The cube of the fifth [quotient] is subtracted from this; remainder: 6, 21, 35, 9, 24, 48, 56. This is added to the remainder: 1, 25, 32, 22, 1, 39, 50, 52. Again, this is divided by the divisor *yā* 3. Quo.: 28, remainder: 1, 32, 22, 1, 39, 50, 52. The quotient is set below the previous quotient: 2, 5, 39, 26, 22, 29, 28.

Its cube: 9, 11, 8, 19, 29, 8, 50, 27, 19, 59, 43. The cube of the sixth [quotient] is subtracted from this; remainder: 6, 8, 25, 40, 8, 51. Added to the remainder: 1, 38, 30, 27, 19, 59, 43. Again divided by the divisor. Quo.: 32, rem.: 2, 30, 27, 19, 59, 43. The quotient is set below the previous quotient: 2, 5, 39, 26, 22, 29, 28, 32.

41 म inserted above line J_1

47 पुनलब्धिरधस्थिता J_2
48 इदं overwritten with अयं J_2
49 योजितंः J_2
50 ० धस्थापितः J_2
51 य overwritten with ज in 2nd hand? J_2

अस्य घनः ९ । ११ । ८। १९ । २९ । ८ । ५७ । २८ । २३ । ३७ । १। अत्र सप्तमघनः शोधितः शेषं ७ । १ । ३ । ३७ । १८। इदं शेषे योजितं २ । ३७ । २८ । २३ । ३७ । १। |[p. 1067 Js] हरभक्ते लब्धं ५२ शेषं १ । २८ । २३ । ३७ । १। लब्धिः पूर्वलब्धेरधः स्थापिता २ । ५ । ३९ । २६ । २२ । २९ । २८ । ३२ । ५२।

अस्य घनः ९ । ११ । ८ । १९ । २९ । ८ । ५७ । ३९ । ४७ । ५० । १९। पुनरत्राष्टमघनः शोधितः शेषं ११ । २४ । १३ । १८। शेषे योजितं १ । ३९ । ४७ । ५० । १९। हरभक्तः। लब्धं ३३ शेषं ० । ४७ । ५० । १९। ल॰पूर्ववत् २ । ५ । ३९ । २६ । २२ । २९ । २८ । ३२ । ५२ । ३३। एतत्पर्यन्तं[42] गृहीतम्।

Its cube: 9, 11, 8, 19, 29, 8, 57, 28, 23, 37, 1. The cube of the seventh [quotient] is subtracted from this; remainder: 7, 1, 3, 37, 18. This is added to the remainder: 2, 37, 28, 23, 37, 1. When divided by the divisor, quotient: 52, remainder: 1, 28, 23, 37, 1. The quotient is set below the previous quotient: 2, 5, 39, 26, 22, 29, 28, 32, 52.

Its cube: 9, 11, 8, 19, 29, 8, 57, 39, 47, 50, 19. Again, the cube of the eighth [quotient] is subtracted from this; remainder: 11, 24, 13, 18. Added to the remainder: 1, 39, 47, 50, 19. Divided by the divisor. Quotient: 33, remainder: 0, 47, 50, 19. Quo. as before: 2, 5, 39, 26, 22, 29, 28, 32, 52, 33, taken up to this [place].

अस्य घनः ९ । ११ । ८ । १९ । २९ । ८ । ५७ । २८ । २३ । ३७ । १ । अत्र सप्तमघनः शोधितः शेषं ७ । १ । ३ । ३७ । १८। इदं शेषे योजितं २ । ३७ । २८ । २३ । ३७ । १। हरभक्तः। ल॰५२ शे॰१ । २८ । २३ । ३७ । १। ल पू॰ल॰ २ । ५ । ३९ । २६ । २२ । २९ । २८ । ३२ । ५२।

अस्य घनः ९ । ११ । ८ । १९ । २९ । ८ । ५७ । ३९ । ४७ । ५० । १९। पुनरत्राष्टमघनः शोधितः शेषं ११ । २४ । १३ । १८। शेषे योजितं १ । ३९ । ४७ । ५० । १९। हरभक्तः। ल॰३३ शे ० । ४७ । ५० । १९। फलं[52] पूर्ववत् २ । ५ । ३९ । २६ । २२ । २९ । २८ । ३२ । ५२ । ३३। एतत्पर्यन्तं गृहीतम् ॥ ॥

Its cube: 9, 11, 8, 19, 29, 8, 57, 28, 23, 37, 1. The cube of the seventh [quotient] is subtracted from this; remainder: 7, 1, 3, 37, 18. This is added to the remainder: 2, 37, 28, 23, 37, 1. Divided by the divisor. Quo.: 52, rem.: 1, 28, 23, 37, 1. Quo. [set below] prev. quo.: 2, 5, 39, 26, 22, 29, 28, 32, 52.

Its cube: 9, 11, 8, 19, 29, 8, 57, 39, 47, 50, 19. Again, the cube of the eighth [quotient] is subtracted from this; remainder: 11, 24, 13, 18. Added to the remainder: 1, 39, 47, 50, 19. Divided by the divisor. Quo.: 33, rem.: 0, 47, 50, 19. Result as before: 2, 5, 39, 26, 22, 29, 28, 32, 52, 33, taken up to this [place].

[42] ए inserted from right margin J_1

[52] फलः J_2

2.4 [Ābida's Successive-Values Iteration for the Chord of 2°]

अथात्र यावत्तावन्मानानयनार्थमुपायान्तरमाबिदेन[43] निष्कासितम्। तद्यथा।

रूपाणि यावत्तावता भाज्यानि। लब्धेर्घनः कार्यः पुनर्घनो ऽपि यावत्तावता भाज्यः। यल्लब्धं तत्प्रथमलब्धौ योज्यं पुनस्तस्य घनः कार्यः। एवमसकृत्।

अत्रोपपत्तिः। इह पूर्ववद्यावत्तावता भक्ते रूपराशौ कश्चिद्यावत्तावतो भागो लब्धः। पुनस्तस्य घनं कृत्वा तद्योजनेन वास्तवघनस्यासन्नता[44] जाता। एवं मुहुः स्थिरीभूतं तदेवघनस्वरूपराशेर्यावत्तावतो भाग उपलब्धः स एव यावत्तावन्मानमुपपन्नम्।

Now here for the sake of the determination of the amount of the *yāvattāvat*, another approach is laid out by Ābida. That is as [follows]:

The units are divided by the *yāvattāvat* [coefficient]. The cube of the quotient is to be made; moreover, the cube too is to be divided by the *yāvattāvat* [coefficient]. Whatever is the quotient is to be added to the first quotient; again, the cube of that is to be made. And so repeatedly.

Here is the demonstration. In this case as before, when the quantity of units is divided by the *yāvattāvat* [coefficient], some part of the *yāvattāvat* is obtained. [When one has] again made the cube of that, by adding [the cube] to that [first quotient and cubing it], an approximation of the cube of the genuine [value] is produced. [When it is carried out] in this way repeatedly to [the state of] fixedness, the part of the *yāvattāvat* achieved from that very quantity of the cube [and] its own units is itself the obtained amount of the *yāvattāvat*.

[43] ∘माविदेन J₁
[44] वास्तवघनस्यांसन्नता J₁

अथात्र यावत्तावन्मानानयनार्थमुपायान्तरमाबिदेन[53] निष्कासितम्[54] ॥ तद्यथा ॥

रूपाणि यावत्तावता भाज्यानि। लब्धेर्घनः कार्यः ॥ पुनर्घनो ऽपि यावत्तावता भाज्यः। यल्लभ्यते तत्प्रथमलब्धौ योज्यं पुनस्तस्य घनः कार्यः ॥ पुनः स[55] यावत्तावता भाज्यः। यल्लब्धं तत्प्रथमलब्धौ योज्यं पुनस्तस्य घनः कार्यः ॥ एवमसकृत् ॥

अत्रोपपत्तिः ॥ इह पूर्ववद्यावत्तावता भक्ते रूपराशौ कश्चिद्यावत्तावतो भागो लब्धः। पुनस्तस्य घनं कृत्वा तद्योजनेन वास्तवघनस्यासन्नता जाता। एवं मूहूः स्थिरीभूतं तदेवावास्तवघनसरूपराशेर्यावत्तावतो[56] भाग उपलब्धः स एव यावत्तावन्मानम् । इत्युपपन्नम्[57] ॥

Now here for the sake of the determination of the amount of the *yāvattāvat*, another approach is laid out by Ābida. That is as [follows]:

The units are divided by the *yāvattāvat* [coefficient]. The cube of the quotient is to be made. Moreover, the cube too is to be divided by the *yāvattāvat* [coefficient]. Whatever is obtained is to be added to the first quotient; again, the cube of that is to be made. Again that is to be divided by the *yāvattāvat* [coefficient]. Whatever is the quotient is to be added to the first quotient; again, the cube of that is to be made. And so repeatedly.

Here is the demonstration. In this case as before, when the quantity of units is divided by the *yāvattāvat* [coefficient], some part of the *yāvattāvat* is obtained. [When one has] again made the cube of that, by adding [the cube] to that [first quotient and cubing it], an approximation of the cube of the genuine [value] is produced. [When it is carried out] in this way repeatedly to [the state of] fixedness, the part of the *yāvattāvat* achieved from that very quantity of the inaccurate cube plus units is itself the amount of the *yāvattāvat*. Thus it is obtained.

[53] ∘पायांतरं आविदेन J₂
[54] निष्काशितं J₂
[55] पुन सः J₂
[56] ∘घनरू∘ with स inserted from right margin J₂
[57] ∘नमित्युपपन्नं with ि crossed out and इ inserted from right margin J₂

अत्रोदाहरणं रू ६ । १६। ४९ । ७ । ५९ । ८ । ५६ । ३०। इदं[45] यावत्तावता[46] षष्टेर्द्विरूर्ध्व|[f. 6r J₁] परिवर्तेन या ३ भक्ते यल्लब्धं फलं २ । ५ । ३६ । २२ । ३९ । ४२। ५८ । ५०। अस्य घनः[47] ९ । १० । २८ । ३ । ८ । ५२ । ५ । ३९। पुनरयं तेनैव या ३ भक्तः फलं ० । ० । ३ । ३ । २९ । २१ । २ । ५७। इदं प्रथमलब्धौ |[p. 1068 J$_S$] योजितं २ । ५ । ३९ । २६ । ९ । ४ । १ । ४७। अस्य घनो वारद्वयं षष्ट्या भक्तः। जातो विकलात्मकः ० । ० । ९ । ११ । ८ । १६। ३२ । ३० । ४८ । ९। पुनरयं तेनैव भक्तः। लब्धं ० । ० । ३ । ३ । ४२ । ४५ । ३० । ५०। इदं प्रथमलब्धौ योजितं जातं ५ । ३९ । २६ । २२ । २८। २९ । ४०। अस्य घनः ९ । ११ । ८ । १९ । २८ । ३६ । २३ । ५०। पुनस्तेनैव भक्तः। लब्धं ० । ० । ३ । ३ । ४२ । ४६ । २९ । ३९। प्रथमलब्धौ योजितं २ । ५ । ३९ । २६ । २२ । २९ । २८ । २९।

अस्य घनः ९ । ११। ८ । १९ । २९ । ८ । ५६ । ५१। पुनस्तेनैव भक्तः। लब्धं ० । ० । ३ । ३। ४२ । ४६ । २९ । ४२। इदं प्रथमलब्धौ योजितं २ । ५ । ३९ । २६। २२ । २९ । २८ । ३२। अस्य घनः ९ । ११ । ८ । १९ । २९ । ८ । ५७ । २८। पुनस्तेनैव भक्तः। लब्धं[48] ० । ० । ३ । ३ । ४२ । ४६ । २९ । ४२। प्रथमलब्धौ योजितं २ । ५ । ३९ । २६ । २२ । २९ । २८ । ३२। अयं स्थिरीभूतः। इयमंशद्वयस्य पूर्णज्या जाता।

Here, an example: *rū* 6, 16, 49, 7, 59, 8, 56, 30. This, when divided by *yā* 3, the *yāvattāvat* [coefficient] twice raised by sixty: whatever is the quotient is the result, 2, 5, 36, 22, 39, 42, 58, 50. Its cube: 9, 10, 28, 3, 8, 52, 5, 39. Again, this is divided by that same *yā* 3; result, 0, 0, 3, 3, 29, 21, 2, 57. This is added to the first quotient: 2, 5, 39, 26, 9, 4, 1, 47. Its cube is twice divided by sixty, resulting as seconds: 0, 0, 9, 11, 8, 16, 32, 30, 48, 9. Again, this is divided by that same [divisor]; quotient, 0, 0, 3, 3, 42, 45, 30, 50. This is added to the first quotient, producing [2,] 5, 39, 26, 22, 28, 29, 40. Its cube: 9, 11, 8, 19, 28, 36, 23, 50. Again divided by that same [divisor]; quotient, 0, 0, 3, 3, 42, 46, 29, 39. Added to the first quotient: 2, 5, 39, 26, 22, 29, 28, 29.

Its cube: 9, 11, 8, 19, 29, 8, 56, 51. Again divided by that same [divisor]; quotient, 0, 0, 3, 3, 42, 46, 29, 42. This is added to the first quotient: 2, 5, 39, 26, 22, 29, 28, 32. Its cube: 9, 11, 8, 19, 29, 8, 57, 28. Again divided by that same [divisor]; quotient, 0, 0, 3, 3, 42, 46, 29, 42. Added to the first quotient: 2, 5, 39, 26, 22, 29, 28, 32. That [has] become fixed. This is the resulting chord of two degrees.

अत्रोदाहरणम् ॥ रू ६ । १६ । ४९ । ७ । ५९ । ८ । ५६ । ३०। इदं यावत्तावता[58] षष्टेर्द्विरूर्ध्वपरिवर्तेन[59] या ३ भक्ते ल॰२ । ५ । ३६ । २२ । ३९ । ४२ । ५८ । ५०। अस्य घनः ९ । १० । २८ । ३ । ८ । ५२ । ५ । ३९। पुनरयं तेनैव या ३ भक्तः[60] फलं वारद्वयं षष्ट्योर्ध्वीकृतं[61] द्विपरिवर्ते योजनार्थं[62] ० । ० । ३ । ३ । २९ । २१ । २ । ५७। इदं प्रथमलब्धौ योजितं २ । ५ । ३९ । २६ । ९ । ४ । १ । ४७। अस्य घनः ९ । ११ । ८ । १६ । ३२ । ३० । ४८ । ९। पुनरयं तेनैव भक्तः। लब्धं ० । ० । ३ । ३ । ४२ । ४५ । ३० । ५०। इदं प्रथमलब्धौ योजितं जातं ५ । ३९ । २६ । २२ । २८ । २९ । ४०। अस्य घनः ९ । ११ । ८ । १९ । २८ । ३६ । २३ । ५०। पुनस्तेनैव भक्तः। लब्धं ० । ० । ३ । ३ । ४२ । ४६ । २९ । ३९। प्रथमलब्धौ योजितं २ । ५ । ३९ । २६ । २२ । २९ । २८ । २९।

अस्य घनः ९ । ११ । ८ । १९ । २९ । ८ । ५६ । ५१। पुनस्तेनैव या ३ भक्तः फ॰० । ० । ३ । ३ । ४२ । ४६ । २९ । ४२। इदं प्रथमलब्धौ योजितं २ । ५ । ३९ । २६ । २२ । २९ । २८ । ३२ । अस्य घनः ९ । ११ । ८ । १९ । २९ । ८ । ५७ । २८। पुनस्तेनैव भक्तः। लब्धं ० । ० । ३ । ३ । ४२ । ४६ । २९ । ४२। प्र॰ल॰योजितः २ । ५ । ३९ । २६ । २२ । २९ । २८ । ३२। अयं स्थिरीभूतः ॥ इयमंशद्वयस्य पूर्णज्या जाता ॥

Here, an example: *rū* 6, 16, 49, 7, 59, 8, 56, 30. This, when divided by *yā* 3, the *yāvattāvat* [coefficient] twice raised by sixty: quo., 2, 5, 36, 22, 39, 42, 58, 50. Its cube: 9, 10, 28, 3, 8, 52, 5, 39. Again, this is divided by that same *yā* 3; the result is twice sixty-raised when twice-shifted for the purpose of addition: 0, 0, 3, 3, 29, 21, 2, 57. This is added to the first quotient: 2, 5, 39, 26, 9, 4, 1, 47. Its cube: 9, 11, 8, 16, 32, 30, 48, 9. Again, this is divided by that same [divisor]; quotient, 0, 0, 3, 3, 42, 45, 30, 50. This is added to the first quotient, producing [2,] 5, 39, 26, 22, 28, 29, 40. Its cube: 9, 11, 8, 19, 28, 36, 23, 50. Again divided by that same [divisor]; quotient, 0, 0, 3, 3, 42, 46, 29, 39. Added to the first quotient: 2, 5, 39, 26, 22, 29, 28, 29.

Its cube: 9, 11, 8, 19, 29, 8, 56, 51. Again divided by that same *yā* 3; res., 0, 0, 3, 3, 42, 46, 29, 42. This is added to the first quotient: 2, 5, 39, 26, 22, 29, 28, 32. Its cube: 9, 11, 8, 19, 29, 8, 57, 28. Again divided by that same [divisor]; quotient, 0, 0, 3, 3, 42, 46, 29, 42. Added 1st quo.: 2, 5, 39, 26, 22, 29, 28, 32. That [has] become fixed. This is the resulting chord of two degrees.

45 यं written over दं and inserted from bottom margin J₁
46 यावत्तावत् J₁
47 घनं J₁
48 वध्बं J₁

58 यावत्ता J₂
59 षष्टेद्वि॰ J₂
60 भक्त J₂
61 ॰कृतः J₂
62 योजिनार्थं J₂

2.5 [Ulugh Beg's Successive-Values Iteration for the Chord of 2°]

अथ मिर्जोलुग्वेगोक्तप्रकारेणांशद्वयस्य पूर्णज्या निष्काश्यते। तत्र पूर्वोक्तमेव क्षेत्रं कार्यम्।

तत्र बचिह्नात् बहलम्बः कजरेखायां निष्काश्यः। तत्र कबहत्रिभुजे जबहत्रिभुजे हकोणः समकोणो ऽस्ति। कबवर्गः कहहब| [f. 6v J_1] योर्वर्गयोगतुल्यो ऽस्ति। पुनर्बजवर्गः जहहबयोर्वर्गयोगतुल्यो ऽस्ति। पुनरत्रकबबजौ मिथस्तुल्यौ कल्पितौ। बहमुभयोस्त्रिभुजयोरेकमेवास्ति। तस्मात्कहहजौ मिथः समानौ भविष्यतः। कबवर्गः बहस्य व्यासस्य च घातेन समानो ऽस्ति। तस्माद्यदि कबवर्गो व्यासेन भाज्यते तदा लब्धं बहप्रमाणं भवति। पुनर्यदि बहवर्गः कबवर्गाच्छोध्यते[49] तदा शेषं कहवर्गो ऽवशिष्यते।

Now by the method stated by Mirjolugvega [Mīrzā Ulugh Beg], the chord of two degrees is expounded. There, that previously stated figure is to be made [fig. J_1.3].

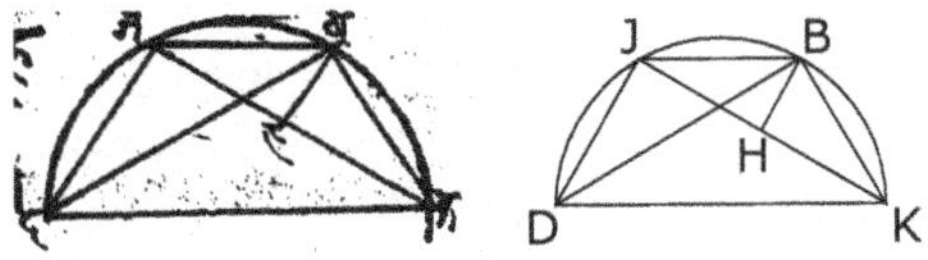

Figure J_1.3: Diagram on f. 6r of MS J_1 (facsimile and reconstruction); *Siddhāntasamrāṭ* ch. 1 fig. 3

In it, from point *B* perpendicular *BH* is extended onto line *KJ*. There in triangle *KBH* [and] in triangle *JBH* angle *H* is a right angle. The square of *KB* is equal to the sum of the squares of *KH* [and] *HB*. Moreover, the square of *BJ* is equal to the sum of the squares of *JH* [and] *HB*. Here again, *KB* [and] *BJ* are considered equal to each other. The same [side] *BH* is in both triangles. Therefore *KH* [and] *HJ* will be equal to each other. The square of *KB* is equal to the product of *BH* and the diameter. Therefore if the square of *KB* is divided by the diameter, then the quotient is the amount of *BH*. Further, if the square of *BH* is subtracted from the square of *KB*, then the remainder, the square of *KH*, is left over.

49 ∘कववर्गाछोध्यते J_1

अथ मिर्योलुग्वेगोक्तप्रकारेणांशद्वयस्य पूर्णज्या निष्काश्यते। तत्र पूर्वोक्तमेवक्षेत्रं कार्यम्।

तत्र बचिह्नाद्बहलम्बः अजरेखायां निष्काश्यते[63]। तत्र अबहत्रिभुजे जबहत्रिभुजे हकोणः समकोणो ऽस्ति ॥ अबवर्गः अहहबयोर्वर्गयोगतुल्यो[64] ऽस्ति ॥ पुनर्बजवर्गः जहहबयोर्वर्गयोगतुल्यो ऽस्ति ॥ पुनरत्र| [f. 3v J_2] अबबजौ मिथस्तुल्यौ कल्पितौ ॥ बहमुभयोस्त्रिभुजयोरेकमेवास्ति। तस्मात्[65] अहहजौ मिथः समानौ[66] भविष्यतः ॥ अबवर्गः। इदं रेखागणितस्य षष्टाध्यायस्य क्षेत्रं ८। १७ मिलित्वोपपद्यते[67]। बहस्य व्यासस्य च घातेन समानो ऽस्ति ॥ तस्माद्यद्याबवर्गो व्यासेन भाज्यते तदा लब्धं बहप्रमाणं[68] भवति ॥ पुनर्यदि बहवर्गो अबवर्गाच्छोध्यते तदा शेषं अहवर्गो ऽवशिष्यते ॥

Now by the method stated by Miryolugvega [Mīrzā Ulugh Beg], the chord of two degrees is expounded. There, that previously stated figure is to be made [fig. J_2.2].

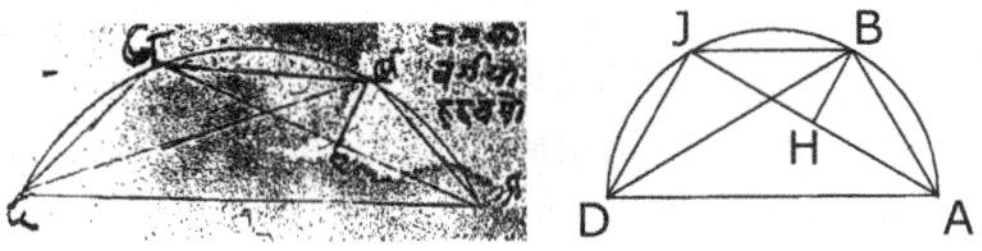

Figure J_2.2: Diagram on f. 3r of MS J_2 (facsimile and reconstruction)

In it, from point *B* perpendicular *BH* is extended onto line *AJ*. There in triangle *ABH* [and] in triangle *JBH* angle *H* is a right angle. The square of *AB* is equal to the sum of the squares of *AH* [and] *HB*. Moreover, the square of *BJ* is equal to the sum of the squares of *JH* [and] *HB*. Here again, *AB* [and] *BJ* are considered equal to each other. The same [side] *BH* is in both triangles. Therefore *AH* [and] *HJ* will be equal to each other. The square of *AB* (this is demonstrated on encountering figures 8 [and] 17 of the sixth chapter of *Rekhāgaṇita*) is equal to the product of *BH* and the diameter. Therefore if the square of *AB* is divided by the diameter, then the quotient is the amount of *BH*. Further, if the square of *BH* is subtracted from the square of *AB*, then the remainder, the square of *AH*, is left over.

63 तत्र∘...∘निष्काश्य inserted from right margin; in it, ∘लम्ब अ∘ partly obscured; J_2

64 अहहवयोवर्गतुल्यो J_2

65 त् inserted from left margin J_2

66 समोनौ J_2

67 इदं∘...∘पद्यते inserted from top margin J_2

68 ∘प्रमाणः J_2

अत्रांश| [p. 1069 J_S] द्वयस्य पूर्णज्यारूपा कबरेखा यावत्तावन्मिता कल्पिता या १। अस्य वर्गः याव १। व्यासः १२० षष्ट्योर्ध्वीकृतः २ एकपरिवर्तः। अनेन कबवर्गे[50] याव १ षष्टेरेकपरिवर्तेन २ भक्ते लब्धं यावद्वर्गादर्धकला सा च यावद्वर्गस्य त्रिंशद्विकलात्मिका ३०। यावत्तावतोंऽशात्मकत्वादिदं बहप्रमाणं जातम्। अस्य वर्गो यावद्वर्गवर्गस्य[51] पञ्चदशप्रतिविकला जाताः १५। इदं कबवर्गाच्छोधितं शेषं कहवर्गः।

याव १ यावव १५ प्रतिविकला ऋ[52]

कहवर्गः कजवर्गस्य चनुर्थांशतुल्यो[53] ऽस्ति। तस्मात्कजवर्गः

याव ४ यावव १̇[54] विकला।

Here the line KB in the form of the chord of two degrees is considered equivalent to the *yāvattāvat*, *yā* 1. Its square is *yāva* 1. The diameter 120, sixty-raised: 2, once-shifted. When the square of KB, *yāva* 1, is divided by that (2 once-shifted by sixty), the quotient is a half-minute with respect to the square of the *yāvattāvat*, and that consists of thirty seconds of the square of the *yāvattāvat*: 30. Because of the *yāvattāvat*'s having the nature of degrees, this amount for BH is produced. Its square is fifteen resulting thirds of the square of the square of the *yāvattāvat*, 15. This is subtracted from the square of KB, [and] the remainder is the square of KH:

yāva 1, *yāvava* −15 thirds.

The square of KH is equal to a fourth part of the square of KJ. Therefore the square of KJ is *yāva* 4, *yāvava* −1 second.

अत्रांशद्वयस्य पूर्णज्यारूपा अबरेखा यावत्तावन्मिता कल्पिता ॥ या १। अस्या वर्गः याव १। व्यासः १२०। षष्ट्योर्ध्वीकृतः[69] २ एकपरिवर्तः। अनेन अबवर्गे याव १ षष्टेरेकपरिवर्तेन २ भक्ते लब्धं यावद्वर्गादर्धकला लब्धा सा च यावद्वर्गस्य[70] त्रिंशद्विकलात्मिकाः ३०। यावत्तावतोंऽशात्मकत्वादिदं बहप्रमाणं जातम् ॥ अस्य वर्गो यावद्वर्गवर्गस्य पञ्चदशप्रतिविकला[71] जाताः १५। इदं अबवर्गाच्छोधितम्। शेषं अहवर्गः।

याव १ यावव ० १५ प्र०वि ऋण

अहवर्गः अजवर्गस्य चथुर्थांशतुल्यो ऽस्ति ॥ तस्मात् अजवर्गः

याव ४ यावव १̇ विकला।

Here the line AB in the form of the chord of two degrees is considered equivalent to the *yāvattāvat*: *yā* 1. Its square is *yāva* 1. The diameter 120, sixty-raised: 2, once-shifted. When the square of AB, *yāva* 1, is divided by that (2 once-shifted by sixty), the quotient is a half-minute with respect to the square of the *yāvattāvat*, and that is thirty [units] of the form seconds of the square of the *yāvattāvat*: 30. Because of the *yāvattāvat*'s having the nature of degrees, this amount for BH is produced. Its square is fifteen resulting thirds of the square of the square of the *yāvattāvat*, 15. This is subtracted from the square of AB. The remainder is the square of AH:

yāva 1, *yāvava* −0, 15‴

The square of AH is equal to a fourth part of the square of AJ. Therefore the square of AJ is *yāva* 4, *yāvava* −1 second.

50 कबवर्गो J_1

51 वर्ग inserted from right margin J_1

52 This boxed line up to प्रतिवि written in thinner script J_1

53 चतुर्थांशत्तुल्यो J_1

54 १ J_1

69 षष्ट्यार्ध्वीकृतः J_2

70 व inserted from right margin J_2

71 पंचदर्शप्रतिविकलाः J_2

अथ मिजस्तीग्रन्थस्य प्रथमाध्यायस्य[55] द्वितीयक्षेत्र इदमुपपन्नम्। कजबदघातः कजवर्गरूपः कबजदघा|[f. 7r J_1] तस्य बजकदघातस्य च योगेन तुल्यो ऽस्ति। तत्र कदप्रमाणं ६। १६। ४९। ७। ५९। ८। ५६। ३०। बजं कबतुल्यं[56] यावन्मितमस्ति। तस्माद्बजं कदेन गुणितं सदेतावन्ति यावत्तावतानि जातानि या ६। १६। ४९। ७। ५९। ८। ५६। ३०। कबवर्गरूपकबजदघातः याव १। तस्मादिदं याव ४ यावव १̇[57] विक० अस्य[58] याव १ या ६। १६। ४९। ७। ५९। ८। ५६। ३० समानं जातम्। पुनरेतौ समशोधितौ। तत्रैकपक्षे।

याव ३
याव०ध १[59] विक० या ६। १६। ४९। ७। ५९। ८। ५६। ३०। एतौ समानौ स्तः।

Now, in the second figure of the first chapter of the book *Mijastī*, this is demonstrated [see figure J_1.2]. The product of *KJ* [and] *BD* has the form of the square of *KJ*, [which] is equal to the sum of the product of *KB* [and] *JD* and the product of *BJ* [and] *KD*. Then the amount of *KD* is 6, 16, 49, 7, 59, 8, 56, 30. *BJ*, equal to *KB*, is the amount of the *yāvattāvat*. Therefore, *BJ* [being] multiplied by *KD*, exactly that many *yāvattāvats* are produced: *yā* 6, 16, 49, 7, 59, 8, 56, 30. The product of *KB* [and] *JD*, having the form of the square of *KB*: *yāva* 1. Therefore this: *yāva* 4 *yāvava* $-1''$ is the resulting equal of this: *yāva* 1, *yā* 6, 16, 49, 7, 59, 8, 56, 30. Again, those two are diminished by equal [amounts]. Then on one side:

yāva 3
yāvava $+1''$, *yā* 6, 16, 49, 7, 59, 8, 56, 30. These two are equal.

अथ मिजिस्तीग्रन्थस्य प्रथमाध्यायस्य[72] द्वितीयक्षेत्र इदमुपपन्नम्। अजबदघातः[73] अजवर्गरूपः अबजदघातस्य बजअदघातस्य च योगेन तुल्यो ऽस्ति ॥ तत्र अदप्रमाणं ६। १६। ४९। ७। ५९[74]। ८। ५६। ३०। बजं अबतुल्यं यावन्मितमस्ति। तस्माद्बजं अदेन गुणितं सदेतावन्ति यावत्तावतानि जातानि[75] ॥ या ६। १६। ४९। ७। ५९। ८। ५६। ३०। अबवर्गरूपअबजदघातः याव १। तस्मादिदं याव ४ यवव १̇ विक० अस्य[76] याव १ या ६। १६। ४९। ७। ५९। ८। ५६। ३० समानं जातम् ॥ पुनरेतौ समशोधितौ। तत्रैकपक्षे।

याव ३
यावव १ विक या ६। १६। ४९। ७। ५९। ८। ५६। ३०

एतौ समानौ स्तः।

Now, in the second figure of the first chapter of the book *Mijistī*, this is demonstrated [see figure J_2.1]. The product of *AJ* [and] *BD* has the form of the square of *AJ*, [which] is equal to the sum of the product of *AB* [and] *JD* and the product of *BJ* [and] *AD*. Then the amount of *AD* is 6, 16, 49, 7, 59, 8, 56, 30. *BJ*, equal to *AB*, is the amount of the *yāvattāvat*. Therefore, *BJ* [being] multiplied by *AD*, exactly that many *yāvattāvats* are produced: *yā* 6, 16, 49, 7, 59, 8, 56, 30. The product of *AB* [and] *JD*, having the form of the square of *AB*: *yāva* 1. Therefore this: *yāva* 4 *yāvava* $-1''$ is the resulting equal of this: *yāva* 1, *yā* 6, 16, 49, 7, 59, 8, 56, 30. Again, those two are diminished by equal [amounts]. Then on one side:

yāva 3
yāvava $1''$, *yā* 6, 16, 49, 7, 59, 8, 56, 30.

These two are equal.

55 प््रथमध्यायस्य J_1
56 अ with erasure mark over it instead of क J_1
57 १ J_1
58 विक० अं अस्य J_1
59 याव ३०...०१ written in thinner script J_1

72 प्रथमध्यायस्य J_2
73 घातोः with vowel stroke struck out J_2
74 ५९ omitted J_2
75 यातानि J_2
76 अ अस्य J_2

अत्र समयोस्त्र्यंशौ चेद्गृह्येते तदा तावपि समानौ भविष्यतः। अत्र पक्षौ त्रिभक्तौ।

याव १
यावव २० प्रतिविकला या २। ५। ३६। २२। ३९। ४२। ५८। ५०[60]

पुनरेतौ यावत्तावतापवर्त्तितौ यथा।

या १
याघ २० रू २। ५। |[p. 1070 J_S] ३६। २२। ३९। ४२। ५८। ५०

तत्र तादृशस्येष्टाङ्कस्य[61] चिकीर्षास्ति। यस्याङ्कस्य रूपराशिस्थिताङ्कस्य चान्तरं तदङ्कघनस्य[62] विंशतिप्रतिविकलात्मकं भवति। स एवाङ्क इष्टो भवति।

Here if two third parts of two equals are taken then those two also will become equal. Here the two sides are divided by three:

yāva 1
yāvava 20 thirds, *yā* 2, 5, 36, 22, 39, 42, 58, 50

Again, those two are reduced by the *yāvattāvat* as [follows]:

yā 1
yāgha 20, *rū* 2, 5, 36, 22, 39, 42, 58, 50

Then [there is] the intention of [finding] such a desired number. The difference of that number and the number in the place of the units quantity consists of twenty thirds of the cube of that number. That is just the number desired.

60 प्रतिविकला inserted below line, प्रति॰...॰५० written in thinner script J_1
61 तत्रादृश॰ J_1
62 तादंकंघनस्य J_1

अत्र समयोस्त्र्यंशौ चेद्गृह्येते[77] तदा तावपि समानौ भविष्यतः। अतः पक्षौ त्रिभक्तौ।

याव १
यावव २० या २। ५। ३६। २२। ३९। ४२। ५८। ५०
प्र॰ वि॰

पुनरेतौ यावत्तावतापवर्तितौ।

या १
याघ २० रू २। ५। ३६। २२। ३९। ४२। ५८। ५०
प्र॰ वि॰

अथात्र तादृशस्येष्टांशस्य चिकीर्षास्ति। यस्याङ्कस्य रूपराशिस्थिताङ्कस्य चान्तरं तदङ्कघनस्य विंशतिप्रतिविकलात्मकमन्तरं भवति। स एवाङ्क इष्टो भवति ॥

Here if two third parts of two equals are taken then those two also will become equal. Here the two sides are divided by three:

yāva 1
yāvava 20‴, *yā* 2, 5, 36, 22, 39, 42, 58, 50

Again, those two are reduced by the *yāvattāvat*:

yā 1
yāgha 20‴, *rū* 2, 5, 36, 22, 39, 42, 58, 50

Now here [there is] the intention of [finding] such a desired number. The difference of that number and the number in the place of the units quantity is a difference consisting of twenty thirds of the cube of that number. That is just the number desired.

77 चद्गृह्यते J_2

तस्योत्पादने ऽयमुपाय उपलब्धः। रूपाणां घनः कृतः ९ । १० । २८ । ३ । ८ । ५२ । ५ । ३९। इदं विंशतिप्रतिविकलाभिर्गुणितं ० । ० । ३ । ३ । २९ । २१ । २ । ५७ रूपेषु योजितं २ । ५ । ३९ । २६ । ९ । ४ । १ । ४७ । अस्य घनः ९ । ११ । ८। १६ । ३२ । ३० । ४८ । ९। पुनरिदं विंशतिप्रतिविकलाभिर्गुणितं ० । ० । ३ । ३ । ४२ । ४५ । ३० । ५० रूपेषु योजितं २ । ५ । ३९ । २६ । २२ । २८ । २९ । ४० । अस्य घनः| [f. 7v J_1] ९ । ११ । ८ । १९ । २८ । ५६ । २३ । ५० पुनस्तेनैव प्र०वि० २० गुणितः ० । ० । ३ । ३ । ४२ । ४६ । २९ । ३९ रूपेषु योजितः २ । ५ । ३९ । २६ । २२ । २९ । २८ । २९।

पुनर्घनः ९ । ११ । ८ । १९ । २९ । ८ । ५६ । ५१ पुनस्तेनैव प्र०वि० २० गुणितः ० । ० । ३ । ३ । ४२ । ४६। २९ । ४२। रूपेषु योजितं २ । ५ । ३९ । २६ । २२ । २९ । २८ । ३२। इदमंशद्वयस्य पूर्णज्यारूप इष्टाङ्को जातः। यदास्य घनः क्रियते विंशतिप्रतिविकलाभिर्गुण्यते[63] रूपेषु योजिते[64] तदा स एवाङ्को भवति।

In its production, this approach is understood. The cube of the units is made: 9, 10, 28, 3, 8, 52, 5, 39. This multiplied by twenty thirds, 0, 0, 3, 3, 29, 21, 2, 57, is added to the units: 2, 5, 39, 26, 9, 4, 1, 47. Its cube: 9, 11, 8, 16, 32, 30, 48, 9. Again, this multiplied by twenty thirds, 0, 0, 3, 3, 42, 45, 30, 50, is added to the units: 2, 5, 39, 26, 22, 28, 29, 40. Its cube, 9, 11, 8, 19, 28, 56, 23, 50, again multiplied by just that 20‴, 0, 0, 3, 3, 42, 46, 29, 39, [and] added to the units: 2, 5, 39, 26, 22, 29, 28, 29.

Again, the cube, 9, 11, 8, 19, 29, 8, 56, 51, again multiplied by just that 20‴: 0, 0, 3, 3, 42, 46, 29, 42. Added to the units: 2, 5, 39, 26, 22, 29, 28, 32. This is the resulting desired number in the form of the chord of two degrees. When its cube is made, [and] is multiplied by twenty thirds, when [that] is added to the units, then it is just that number.

तस्योत्पादने ऽयमुपाय उपलब्धः॥ रूपाणां घनः कृतः। ९ । १० । २८ । ३ । ८ । ५२ । ५ । ३९। इदं विंशतिप्रतिविकलाभिर्गुणितं[78] ० । ० । ३ । ३ । २९ । २१ । २ । ५७ रूपेषु योजितं २ । ५ । ३९ । २६ । ९ । ४ । १ । ४७ । अस्य घनः ९ । ११ । ८ । १६ । ३२ । ३० । ४८ । ९। पुनरिदं विंशतिप्रतिविकलाभिर्गुणितं ० । ० । ३ । ३ । ४२ । ४५ । ३०। ५० रूपेषु योजितं २ । ५ । ३९ । २६ । २२ । २८ । २९ । ४०। अस्य घनः ९ । ११ । ८ । १९ । २८ । ५६ । २३ । ५० पुनस्तेनैव प्र वि० २० गुणितः ० । ० । ३ । ३ । ४२ । ४६ । २९ । ३९। रू० योजितं॥ २ । ५ । ३९ । २६ । २२ । २९ । २८ । २९।

पुनर्घनः ९ । ११ । ८ । १९ । २९ । ८ । ५६ । ५१ पुनस्तेनैव प्र०वि० २० गुणितः ० । ० । ३ । ३ । ४२ । ४६ । २९ । ४२। रू० योजितं २ । ५ । ३९ । २६ । २२ । २९ । २८ । ३२। इदमंशद्वयस्य[79] पूर्णज्यारूपमिष्टाङ्को जातः॥ यदा| [f. 4r J_2] स्य घनः॥ क्रियते विंशतिप्रतिविकलाभिर्गुण्यते रूपेषु योज्यते तदा स एवाङ्को भवति॥

In its production, this approach is understood. The cube of the units is made: 9, 10, 28, 3, 8, 52, 5, 39. This multiplied by twenty thirds, 0, 0, 3, 3, 29, 21, 2, 57, is added to the units: 2, 5, 39, 26, 9, 4, 1, 47. Its cube: 9, 11, 8, 16, 32, 30, 48, 9. Again, this multiplied by twenty thirds, 0, 0, 3, 3, 42, 45, 30, 50, is added to the units: 2, 5, 39, 26, 22, 28, 29, 40. Its cube, 9, 11, 8, 19, 28, 56, 23, 50, is again multiplied by just that 20‴: 0, 0, 3, 3, 42, 46, 29, 39. Added to the u.: 2, 5, 39, 26, 22, 29, 28, 29.

Again, the cube, 9, 11, 8, 19, 29, 8, 56, 51, is again multiplied by just that 20‴: 0, 0, 3, 3, 42, 46, 29, 42. Added to the u.: 2, 5, 39, 26, 22, 29, 28, 32. This is the resulting desired number in the form of the chord of two degrees. When its cube is made, [and] is multiplied by twenty thirds [and] added to the units, then it is just that number.

63 विंशातिप्रतिकला० J_1

64 ज्योजिते J_1

78 ०भिर्गुणितः J_2

79 इदंमंशद्वस्य J_2

2.6 [Ulugh Beg's Successive-Values Iteration for the Sine of 1°]

अथ मिर्जोलुग्वेगस्य द्वितीयः प्रकारः। तत्र कबजदं चापं षडंशानां कल्पं यस्य वृत्तस्येदं चापं तद्वृत्तकेन्द्रं तं कल्पनीयम्। पुनः प्रत्येकं कबचापं बजचापं जदचापमंशद्वयं कल्पितम्। कबकजकदबजबदजदपूर्णज्याः[65] संयोज्याः। पुनः कबपूर्णज्याकजपूर्णज्याकदपूर्णज्यानां[66] हझबचिह्नेष्वर्धं[67] कार्यम्।

पुनस्तहरेखातझतबरेखाः संयोज्याः। एता रेखाः प्रत्येकं तासु पूर्णज्यासु लम्बरूपा भविष्यन्ति। पुनः कतरेखाव्यासार्धं संयोज्यम्। पुनस्तत् |[p. 1071 Js] अचिह्ने ऽर्धितं कार्यम्। पुनः अचिह्नं केन्द्रं कृत्वा अकव्यासार्धेन वृत्तं कार्यम्।

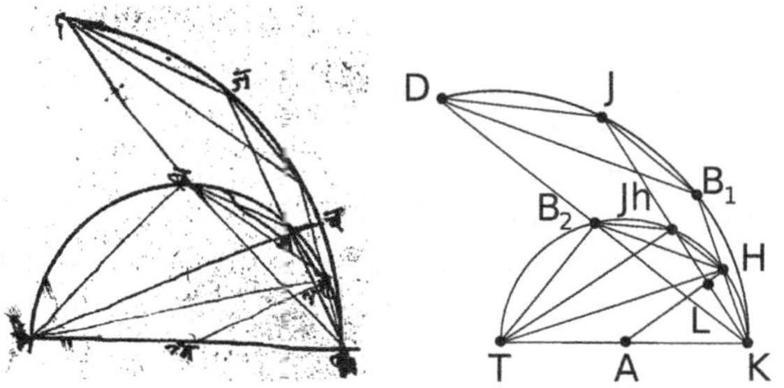

Figure J_1.4 Diagram on f. 8r of MS J_1
(facsimile and reconstruction)

Now, the second method of Mirjolugvega. In it, arc KB_1JD is the equivalent of six degrees (this [is an] arc of that circle whose center is supposed [at] T). Furthermore, each one [of] arc KB_1, arc B_1J, [and] arc JD is considered two degrees. The chords of KB_1, KJ, KD, B_1J, B_1D, [and] JD are to be joined. Furthermore, half of [each of] the chord of KB_1, the chord of KJ, [and] the chord of KD is to be made at the points H, Jh, [and] B_2 [respectively].

Furthermore, line TH [and] lines TJh [and] TB_2 are to be joined. Those lines will each be [in] the form of a perpendicular on those chords. Furthermore, the half-diameter, line KT, is to be joined; furthermore, that is to be made halved at point A. Furthermore, making point A the center, a circle is to be made with half-diameter AK.

अथ मिर्योलुग्वेगस्य द्वितीयप्रकारः ॥ तत्र अबजदचापं षडंशानां कल्प्यम् ॥ यस्य वृत्तस्येदं चापं तद्वृत्तकेन्द्रं तं कल्पनीयम् ॥ पुनः प्रत्येकं अबचापं बजचापं जदचापमंशद्वयं[80] कल्पितम् ॥ अबअजअदबजबदजदपूर्णज्याः[81] संयोज्याः ॥ पुनः अबपूर्णज्याअजपूर्णज्याअदपुर्णज्यानां[82] च[83] हझबचिह्नेष्वर्धं कार्यम् ॥

पुनस्तहरेखातझतबरेखाः संयोज्याः ॥ एता रेखाः प्रत्येकं तासु पूर्णज्यासु लम्बरूपा भविष्यन्ति ॥ पुनः अतरेखाव्यासार्धं संयोज्यम् ॥ पुनः कचिह्ने ऽर्धितं कार्यम्। पुनः कचिह्नं केन्द्रं कृत्वा कअव्यासार्धेन वृत्तार्धं कार्यम् ॥

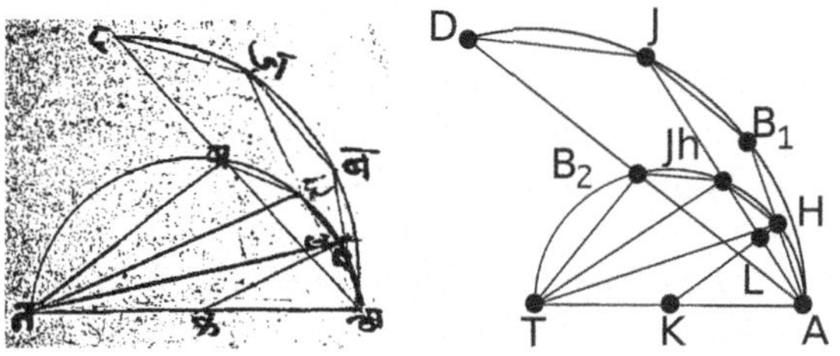

Figure J_2.3: Diagram on f. 4r of MS J_2
(facsimile and reconstruction)

Now, the second method of Miryolugvega. In it, arc AB_1JD is the equivalent of six degrees. This [is an] arc of that circle whose center is supposed [at] T. Furthermore, each one [of] arc AB_1, arc B_1J, [and] arc JD is considered two degrees. The chords of AB_1, AJ, AD, B_1J, B_1D, [and] JD are to be joined. Furthermore, half of [each of] the chord of AB_1, the chord of AJ, and the chord of AD is to be made at the points H, Jh, [and] B_2 [respectively].

Furthermore, line TH [and] lines TJh [and] TB_2 are to be joined. Those lines will each be [in] the form of a perpendicular on those chords. Furthermore, the half-diameter, line AT, is to be joined. Furthermore, that is to be made halved at point K; furthermore, making point K the center, a half-circle is to be made with half-diameter KA.

65 तवकजकदवज॰ J_1
66 ॰ज्याना J_1
67 झँ (?) erased before र्द्धं J_1

80 अचापं वदचापं जदचापमशद्वयं J_2
81 बद omitted J_2
82 नां inserted from top margin J_2
83 चा J_2

तत्र कहतकोणकझतकोणकबतकोणाः समकोणाः सन्ति तदा वृत्तार्धमेतत् हझबचिह्नेषु गमिष्यति। पुनः हझझबबतहतरेखाः संयोज्याः। हचिह्नात् हललम्बः कझपूर्णज्यासु निष्काश्यः। | [f. 8r J_1] तदा कझपूर्णज्या लचिह्नेष्वर्धिता भविष्यति।

तत्र झबरेखया कजभुजकदभुजयोरर्धस्थाने कर्तितं तदा झबरेखा जदरेखायाः समानान्तरा भविष्यति। तस्मात्कजदत्रिभुजकझबत्रिभुजे मिथः सजातीये भविष्यतः[68]। कबं कदस्यार्धमस्ति। झबं जदस्यार्धं भविष्यति। पुनरनेनैव प्रकारेण हझरेखा बजस्यार्धं भविष्यति। कहं कबस्यार्धमस्यैव[69]। तस्मात् कहहझझबपूर्णज्या[70] मिथः समाना भविष्यन्ति।

पुनः अचिह्नात् अललम्बः कझपूर्णज्यायां निष्काश्यः। अयं लम्बः कझरेखां लचिह्ने ऽर्धितां करिष्यति। पुनरयं वर्धितः सन् हचिह्ने लगिष्यति। तस्मात्कहमेकांशज्या यावत्तावन्मिता कल्प्या।

तत्र अहतकोणअझतकोणअबतकोणाः समकोणाः सन्ति ॥ तदा वृत्तार्धमेतत् हझबचिह्नेषु गमिष्यति ॥ पुनः हझझबबतहतरेखाः[84] संयोज्याः ॥ हचिह्नात् हललम्बः[85] अझपूर्णज्यासु निष्काश्यः ॥ तदा अझपूर्णज्या लचिह्नेष्वर्धिता भविष्यति ॥

तत्र झबरेखया अजभुजअदभुजयोरर्धस्थाने कर्तोतम् ॥ तदा झबरेखा जदरेखायाः समानान्तरा भविष्यति ॥ तस्मात् अजदत्रिभुजअझबत्रिभुजे मिथः सजातीये भविष्यतः ॥ अबं अदस्यार्धमस्ति। झबं जदस्यार्धं भविष्यति ॥ पुनरनेनैव प्रकारेण हझरेखा बजस्यार्धं भविष्यति ॥ अहं अबस्यार्धमस्यैव[86] ॥ तस्मात् अहहझझबपूर्णज्या मिथः समानाः भविष्यन्ति ॥

पुनः कचिह्नात् कललम्बः अझपूर्णज्यायां निष्काश्यः ॥ अयं लम्बो अझरेखां। लचिह्ने ऽर्धितां करिष्यति ॥ पुनरयं वर्धितः सन् हचिह्ने लगिष्यति ॥ तस्मात्[87] अहमेकांशज्या यावत्तावन्मिता कल्प्या ॥

In it, angle KHT, angle $KJhT$, [and] angle KB_2T being right angles, then that half circle will go through points H, Jh, [and] B_2. Furthermore, lines HJh, JhB_2, B_2T, [and] HT are to be joined. From point H the perpendicular HL is to be extended on the chords [actually chord, sg.] KJh. Then the chord of KJh will be halved at points [actually point, sg.] L.

There, with line JhB_2 cut[ting] in the half-place of side KJ [and] side KD [i.e., its endpoints Jh and B_2 bisect those two segments], then line JhB_2 will be proportional with respect to line JD. Therefore triangle KJD and triangle $KJhB_2$ will be similar to each other. KB_2 is half of KD; [so] JhB_2 will be half of JD. Furthermore, in just this way line HJh will be half of B_1J. KH is half of KB_1 itself. Therefore the chords KH, HJh, [and] JhB_2 will be equal to one another.

Furthermore, from point A perpendicular AL onto chord KJh is to be extended. This perpendicular will make line KJh halved at point L. Furthermore, this [line], being lengthened, will fall on point H. Therefore KH, the sine of one degree, is to be considered the amount of the *yāvattāvat*.

In it, angle AHT, angle $AJhT$, [and] angle AB_2T being right angles; then that half circle will go through points H, Jh, [and] B_2. Furthermore, lines HJh, JhB_2, B_2T, [and] HT are to be joined. From point H the perpendicular HL is to be extended on the chords [actually chord, sg.] AJh. Then the chord of AJh will be halved at points [actually point, sg.] L.

There, with line JhB_2 cut[ting] in the half-place of side AJ [and] side AD [i.e., its endpoints Jh and B_2 bisect those two segments]: then line JhB_2 will be proportional with respect to line JD. Therefore triangle AJD and triangle $AJhB_2$ will be similar to each other. AB_2 is half of AD; [so] JhB_2 will be half of JD. Furthermore, in just this way line HJh will be half of B_1J. AH is half of AB_1 itself. Therefore the chords AH, HJh, [and] JhB_2 will be equal to one another.

Furthermore, from point K perpendicular KL onto chord AJh is to be extended. This perpendicular will make line AJh halved at point L. Furthermore, this [line], being lengthened, will fall on point H. Therefore AH, the sine of one degree, is to be considered the amount of the *yāvattāvat*.

68 भविष्यंतः J_1

69 ०मस्येव J_1

70 ः inserted above line after ज्या J_1

84 हत inserted from right margin J_2

85 हललव J_2

86 ०मत्स्येव J_2

87 तस्मा J_2

पुनः कबं अंशत्रितयस्य[71] ज्यैतावत्यस्ति ३। ८। २४। ३३। ५९। ३४। २८। १५। यदि कहवर्गः याव १ कतरेखया ६० भाज्यते तदा लब्धिरेका कला यावद्वर्गस्य भविष्यति। इदं हलप्रमाणमस्ति यथा पूर्वमुपपन्नम्। अस्य वर्गः |[p. 1072 J_S] यावद्वर्गवर्गस्यैका| [f. 8v J_1] विकला भविष्यति। अयं कहवर्गे शोधितः शेषं कलवर्गो भविष्यति। स च याव १ यावव ९̇[72] विकला। कझवर्गस्य चतुर्थांशः[73] कलवर्गो ऽस्ति। तस्मात् कझवर्गः याव ४ यावव ४̇ विकलाः[74]।

पुनः कझहबघातरूपकझवर्गः[75] कहझबघातरूपकहवर्गस्य[76] हझकबघातस्य च योगेन तुल्यो ऽस्ति। कझवर्गश्चैतावान्[77] याव १ हझकबघातश्च या[78] ३। ८। २४। ३३। ५९। ३४। २८। १५। अयं द्वितीयप्रकारेण कझवर्गः सिद्धः।

Furthermore, KB_2, the sine of three degrees, is so much: 3; 8, 24, 33, 59, 34, 28, 15. If the square of KH, *yāva* 1, is divided by line *KT* [or] 60, then the quotient will be one minute [sixtieth] of the square of the *yāvattāvat*. This is the amount of HL, as previously obtained. Its square will be one second of the square of the square of the *yāvattāvat*. This is subtracted from the square of KH [and] the remainder will be the square of KL. And that is *yāva* 1, *yāvava* −1 second. A fourth part of the square of KJh is the square of KL. Therefore the square of KJh is *yāva* 4, *yāvava* −4 seconds.

Furthermore, the square of KJh [having] the form of the product of KJh [and] HB_2 is equal to the sum of the square of KH (having the form of the product of KH [and] JhB_2) and the product of HJh [and] KB_2. And the square of KJh is so much: *yāva* 1 and the product of HJh [and] KB_2, 3; 8, 24, 33, 59, 34, 28, 15. This is the square of KJh established by the second method.

पुनः अबमंशत्रितयज्यैतावत्यस्ति ॥ ३। ८। २४। ३३। ५९। ३४। २८। १५। यदि अहवर्गः याव १ अतरेखया ६० भाज्यते तदा लब्धिरेका कला यावद्वर्गस्य भविष्यति। इदं हलप्रमाणमस्ति ॥ यथा पूर्वमुपपन्नम् ॥ अस्यवर्गः यावद्वर्गवर्गस्यैका विकला भविष्यति ॥ अयं अहवर्गे शोधितः शेषं अलवर्गो भविष्यति ॥ स च याव १ यावव ९̇ विकला। अझवर्गस्य चतुर्थांश ४ अलवर्गो ऽस्ति। तस्मात् अझवर्गः याव ४ यावव ४̇ विकलाः।

पुनः अझहबघातरूपअझवर्गः अझबहघातरूपअहवर्गस्य[88] हझअबघातस्य च योगेन तुल्यो ऽस्ति ॥ अझवर्गश्चैतावान्[89] याव १ हझअबघा|[f. 4v J_2] तश्च या ३। ८। २४। ३३। ५९। ३४। २८। १५। अयं द्वितीयप्रकारेणा ऽझवर्गः सिद्धः।

Furthermore, AB_2, the sine of three degrees, is so much: 3; 8, 24, 33, 59, 34, 28, 15. If the square of AH, *yāva* 1, is divided by line *AT* [or] 60, then the quotient will be one minute [sixtieth] of the square of the *yāvattāvat*. This is the amount of HL; as previously obtained. Its square will be one second of the square of the square of the *yāvattāvat*. This is subtracted from the square of AH [and] the remainder will be the square of AL. And that is *yāva* 1, *yāvava* −1 second. A fourth part of the square of AJh is the square of AL. Therefore the square of AJh is *yāva* 4, *yāvava* −4 seconds.

Furthermore, the square of AJh [having] the form of the product of AJh [and] HB_2 is equal to the sum of the square of AH (having the form of the product of AJh [and] B_2H [?]) and the product of HJh [and] AB_2. And the square of AJh is so much: *yāva* 1 and the product of HJh [and] AB_2, 3; 8, 24, 33, 59, 34, 28, 15. This is the square of AJh established by the second method.

[71] ०त्रितयंस्य J_1
[72] ९ J_1
[73] चतुर्थांशः ध J_1
[74] ४ विकला J_1
[75] ०रूपः० J_1
[76] कझहवघातरूपवत् वर्गस्य with २ and ९ over झ and ह respectively? J_1
[77] कहवर्ग० J_1
[78] या omitted J_1

[88] ०रूपवहवर्गस्य J_2
[89] अह० J_2

एतौ समाविति समशोधनार्थं न्यासः। यथा।

याव ४ यावव ४̇ विकलाः[79]
याव १ या ३ । ८ । २४ । ३३ । ५९ । ३४ । २८ । १५

अत्र सम्प्रदायेन शोधने कृते सति प्रथमपक्षे[80]।

याव ३ यावव ४̇ विकलाः
या ३ । ८ । २४ । ३३ । ५९ । ३४ । २८। १५

अनयोस्त्र्यंशावपि समानौ[81]। तस्मात्त्रिभिरपवर्तितौ।
याव १
यावव १ वि० २० प्र०वि० या[82] १ । २ । ४८ । ११ । १९। ५१ । २९ । २५

अवशिष्टावेतावपि समानौ। पुनरेतौ यावत्तावतापवर्तितौ जातौ पक्षौ।
या १
याघ १ वि० २० प्र०वि० रू १ । २ । ४८ । ११ । १९ । ५१ । २९ । २५

"The two are equal": the layout for the sake of subtraction of equals. As [follows]:
yāva 4, *yāvava* −4″
yāva 1, *yā* 3; 8, 24, 33, 59, 34, 28, 15

Here by established usage, when subtraction is done in the first side:

yāva 3, *yāvava* −4″
yā 3; 8, 24, 33, 59, 34, 28, 15

Two third parts of those two are also equal. Therefore the two [sides] are reduced by three:

yāva 1
yāvava 1″, 20‴, *yā* 1; 2, 48, 11, 19, 51, 29, 25

These two remaining [sides] also are equal. Again, those two are reduced by the *yāvattāvat* [and] the two resulting sides [are]:

yā 1
yāgha 1″, 20‴, *rū* 1; 2, 48, 11, 19, 51, 29, 25

एतौ समाविति शोधनार्थं न्यासः।

याव ४ यावव ४̇ विकलाः
याव १ या ३ । ८ । २४ । ३३ । ५९ । ३४ । २८ । १५
अत्र यावनीयसम्प्रदायेन शोधने कृते प्रथमपक्षे याव ३। द्वितीयपक्षे यावव ४̇[90] विकला या ३ । ८ । २४ । ३३ । ५९ । ३४ । २८ । १५।

अनयोस्त्र्यंशावपि[91] समानौ। तस्मात्त्रिभिरपवर्तितौ

याव १
यावव १ वि० २० प्र०वि० या १ । २ । ४८ । ११ । १९ । ५१ । २९ । २५

अवशिष्टावेतावपि समानौ। पुनरेतौ यावत्तावतापवर्तितौ जातौ।
या १
याघ १ वि० २० प्र०वि० रू० १ । २ । ४८ । ११ । १९ । ५१ । २९ । २५

"The two are equal": the layout for the sake of subtraction [of equals]:
yāva 4, *yāvava* −4″
yāva 1, *yā* 3; 8, 24, 33, 59, 34, 28, 15

Here by Yāvanī established usage, when subtraction is done, on the first side: *yāva* 3. On the second side: *yāvava* −4″, *yā* 3; 8, 24, 33, 59, 34, 28, 15.

Two third parts of those two are also equal. Therefore the two [sides] are reduced by three:

yāva 1
yāvava 1″, 20‴, *yā* 1; 2, 48, 11, 19, 51, 29, 25

These two remaining [sides] also are equal. Again, those two are reduced by the *yāvattāvat* [and] the two resulting [sides are]:

yā 1
yāgha 1″, 20‴, *rū* 1; 2, 48, 11, 19, 51, 29, 25

79 ४ विवला J_1
80 धने कृते सति प्रथमपक्षे written twice, with the first occurrence struck out J_1
81 समारौ J_1
82 २० प्र०वि० omitted, रू instead of या J_1

90 ४ J_2
91 अनयो अंशा० J_2

यद्यत्र तादृशो ऽङ्को[83] लभ्यते यस्याङ्कस्य पुनो रूपस्य चान्तरमङ्कघनस्यैकविकलाविंशतिप्रतिविकलातुल्यं[84] भवेत्। स चानेन प्रकारेण निष्काशितः। तत्र रूपाणां घनोंऽशादि १। ८। ४८। ३०। २३। ३६। ३०। ४२। १८। अयमेकया विकलया[85] विंशतिप्रतिवि|[f. 9r J_1] कलाभिश्च गुणितः ०। ०। १। ३१। ४४। ४०। ३१[86]। २८। ४१। अंशकलास्थानयोरभावात्स्थानद्वये शून्यं निवेशितं रूपेषु संयोज्य जातं १। २। ४९। ४३। ४। ३२। ०। ५३। ४१। अस्य घनः १। ८। ५३। ३२। ४। ३। |[p. 1073 J_S] ५०। ५९। १४। ५७। पुनरयमेतेन $\begin{smallmatrix}१\\२०\end{smallmatrix}$ गुणितः ०। ०। १। ३१। ५१। २२। ४५। २५। ८। रूपेषु योजितं १। २। ४९। ४३। ११। १४। १४। ५०। ८।

Here when such a number is obtained, moreover, the difference of that number and the units is equal to one second and twenty thirds of the cube of the number. And it is laid out by this method: There the cube of the units beginning with degrees [is] 1, 8, 48, 30, 23, 36, 30, 42, 18. This is multiplied by one second and twenty thirds: 0, 0, 1, 31, 44, 40, 31, 28, 41. Having added to the units [that result with] zero entered in the [first] two places because of the nonexistence [of any number] in the two places of degrees and minutes, result: 1, 2, 49, 43, 4, 32, 0, 53, 41. Its cube: 1, 8, 53, 32, 4, 3, 50, 59, 14, 57. Again, this is multiplied by that 1, 20: 0, 0, 1, 31, 51, 22, 45, 25, 8. Added to the units: 1, 2, 49, 43, 11, 14, 14, 50, 8.

[83] ताद्यशोंको J_1

[84] ०स्यैककला० J_1

[85] कलया J_1

[86] ३ J_1

यद्यत्र तादृशो ऽङ्को लभ्यते यस्याङ्कस्य पुनो रूपस्य चान्तरमङ्कघनस्यैकविकलाविंशतिप्रतिविकलातुल्यं[92] भवेत् ॥ स चानेन प्रकारेण निष्काशितः ॥ तत्र रूपाणां घनोंऽशादि १। ८[93]। ४८। ३०। २३। ३६। ३०। ४२। १८। अयमेकया विकलया[94] विंशतिप्रतिविकलाभिश्च गुणितः ०। ०। १। ३१। ४४। ४०। ३१ । २८। ४१। अंशकलास्थानयोरभावात्स्थानद्वये शून्यं निवेशितं रूपेषु संयोज्य जातं १। २। ४९। ४३। ४। ३२। ०। ५३। ४१। अस्य घनः १। ८। ५३। ३२। ४। ३। ५०। ५९। १४। ५७। पुनरयमेतेन $\begin{smallmatrix}१\\२०\end{smallmatrix}$ गुणितः ०। ०। १। ३१। ५१। २२। ४५। २५। ८। रूपेषु योजितं १। २। ४९। ४३। ११। १४। १४। ५०। ८।

Here when such a number is obtained, moreover, the difference of that number and the units is equal to one second and twenty thirds of the cube of the number. And it is laid out by this method: There the cube of the units beginning with degrees [is] 1, 8, 48, 30, 23, 36, 30, 42, 18. This is multiplied by one second and twenty thirds: 0, 0, 1, 31, 44, 40, 31, 28, 41. Having added to the units [that result with] zero entered in the [first] two places because of the nonexistence [of any number] in the two places of degrees and minutes, result: 1, 2, 49, 43, 4, 32, 0, 53, 41. Its cube: 1, 8, 53, 32, 4, 3, 50, 59, 14, 57. Again, this is multiplied by that 1, 20: 0, 0, 1, 31, 51, 22, 45, 25, 8. Added to the units: 1, 2, 49, 43, 11, 14, 14, 50, 8.

[92] चान्तरं अकघनस्यैककलाविंशतिप्रतिविकलातुल्यमंतरं J_2

[93] ४८ J_2

[94] कलया J_2

अस्य घनः १। ८। ५३। ३६। २६। ७। ०[87]। २२। १०। पुनरयं तेनैव १ २० गुणितः ०। ०। १। ३१। ५१। २३। १४। ४९। २०। रूपेषु संयोज्य १। २। ४९। ४३। ११। १४। ४४। १४। २०[88]। अस्य घनः १। ८। ५३। ३२। २६। ८। ३७। ५। ३४। पुनस्तेनैव १ २० गुणितः ०। ०। १। ३१। ५१। २३। १४। ५१। २९। रूपेषु संयोज्य १। २। ४९। ४३। ११। १४। ४४। १६। २६।

अयमिष्टाङ्को जातः। अत इयमेकांशज्या सिद्धा। कुतः। यतो ऽस्य घने ऽनेन १ वि० २० प्र० गुणिते रूपेषु योज्यते तदायमेव भवति। तस्मादयमेवैकांशज्या जाता। इदमेवेष्टम्।

Its cube: 1, 8, 53, 36, 26, 7, 0, 22, 10. Again, this is multiplied by that 1, 20: 0, 0, 1, 31, 51, 23, 14, 49, 20. Having added [it] to the units: 1, 2, 49, 43, 11, 14, 44, 14, 20. Its cube: 1, 8, 53, 32, 26, 8, 37, 5, 34. Again, this is multiplied by just that 1, 20: 0, 0, 1, 31, 51, 23, 14, 51, 29. Having added [it] to the units: 1, 2, 49, 43, 11, 14, 44, 16, 26.

This is the resulting desired number. Hence this sine of one degree is attained. Wherefore? Because when its cube is multiplied by this 1″ 20‴ [and] added to the units, then [it] is [still] just this. Hence this very [number] is the resulting sine of one degree; just this is the desired [value].

87 ७। १८। ० J1

88 १६। ३० J1

अस्य घनः १। ८। ५३। ३६। २६। ७। ०[95]। २२। १०। पुनरयं तेनैव १ २० गुणितः ॥ ०। ०। १। ३१। ५१। २३। १४। ४९। २०। रूपेषु संयोज्य १। २। ४९। ४३। ११। १४। ४४। १४। २०[96]। अस्य घनः १। ८। ५३। ३२। २६। ८। ३७। ५। ३४। पुनस्तेनैव १ २० गुणितः ०। ०। १। ३१। ५१। २३। १४। ५१। २९। रूपेषु संयोज्य[97] १। २। ४९। ४३। ११। १४। ४४। १६। २६।

अयमिष्टाङ्को जातः[98]। अत इयमेकांशज्या सिद्धा ॥ कुतः ॥ यतो ऽस्य घने ऽनेन १ विक० २० प्रवि०[99] गुण्यते रूपेषु योज्यते तदायमेव भवति ॥ तस्मादयमेवैकांशज्या जाता। इदमेवेष्टम् ॥ ॥

Its cube: 1, 8, 53, 36, 26, 7, 0, 22, 10. Again, this is multiplied by that 1, 20: 0, 0, 1, 31, 51, 23, 14, 49, 20. Having added [it] to the units: 1, 2, 49, 43, 11, 14, 44, 14, 20. Its cube: 1, 8, 53, 32, 26, 8, 37, 5, 34. Again, this is multiplied by just that 1, 20: 0, 0, 1, 31, 51, 23, 14, 51, 29. Having added [it] to the units: 1, 2, 49, 43, 11, 14, 44, 16, 26.

This is the resulting desired number. Hence this sine of one degree is attained. Wherefore? Because when its cube is multiplied by this 1″ 20‴ [and] added to the units, then [it] is [still] just this. Hence this very [number] is the resulting sine of one degree; just this is the desired [value].

95 ७। १८। ० J2

96 १६। ३० J2

97 संयोज्यः J2

98 अयंमिष्टांको जात J2

99 यतोस्थघने अनेन १ २० प्रवि० विक J2

2.7 [Ābida's Alternate Proof of Ulugh Beg's Sine 1° Method]

अथात्र मिर्जोलुग्वेगेन यद्द्वितीयप्रकारेण क्षेत्रं प्रदर्शितं तत्र बहुरेखासंयोगः कृतः। अथात्र यथाल्परेखाभिरे|[f. 9v J_1] व सिध्यति तथायतितमाबिदसंज्ञैः[89]। तत्र तावदेव कबजदचापं पूर्वोत्त्येव पूर्णज्याः संयोज्याः। कबपूर्णज्यार्धरूपं कहं एकांशज्यास्ति। सा यावन्मिता कल्पिता या १ कबरेखा या २। एवं बज या[90] २ जद या २ यत एता मिथः समानाः सन्ति। पुनः कदरेखा ६। १६। ४९। ७। ५९। ८। १६। अनया |[p. 1074 J_S] बजं या २ गुणितं कबजदघातयुतं जातं याव ४ या १२। ३३। ३८। १५। ५८। १६। ३३। ०। अयं जातः कजवर्गः।

पुनः प्रकारान्तरेण कजवर्गः साध्यते। कबवर्गः याव ४। अयं व्यासेन २ षष्टेरेकोर्ध्वपरिवर्तेन भक्तो लब्धं यावद्वर्गस्य कला २। इदमंशद्वयस्योत्क्रमज्या जाता।

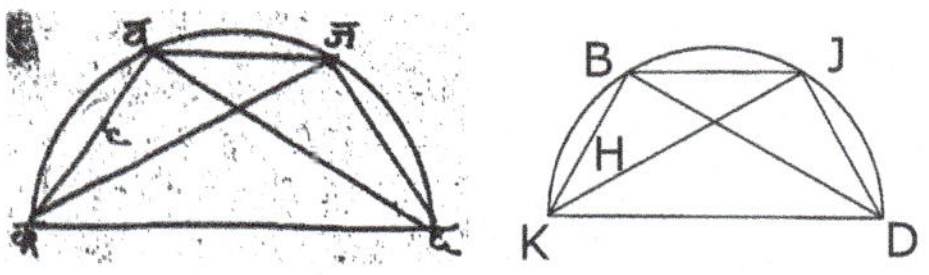

Figure J_1.5: Diagram on f. 9v of MS J_1 (facsimile and reconstruction)

Now here [is] the figure which [was] shown by Mirjolugvega with the second method; in it the connection of many lines is made. Now here it is effected with only a few lines just as it is attained by the one called Ābida. In that [method] just that much [is] arc $KBJD$; [by] the previous statement itself, the chords are to be connected. KH, having the form of half the chord of KB, is the sine of one degree. It is considered the amount of the *yāvattāvat*, *yā* 1, [and] line KB is *yā* 2. In the same way [line] BJ is *yā* 2 [and line] JD is *yā* 2, since they are equal to one another. Furthermore, line KD is 6; 16, 49, 7, 59, 8, 16. BJ [or] *yā* 2 is multiplied by that [and] added to the product of KB [and] JD; result, *yāva* 4, *yā* 12; 33, 38, 15, 58, 16, 33, 0. This result is the square of KJ.

Furthermore, the square of KJ is to be determined by another method. The square of KB is *yāva* 4. This is divided by the diameter, 2 once-raised by sixty; the quotient is 2 minutes of the square of the *yāvattāvat*. This is the resulting versine of two degrees.

89 ०माविदसं० J_1

90 य J_1

अथात्र मिर्योलुग्वेगेन यद्द्वितीयप्रकारेण क्षेत्रं[100] दर्शितं तत्र बहुरेखासंयोगः कृतः ॥ अथात्र यथाल्परेखाभिरेव सिध्यति तथायतितं महमदआबिदसंज्ञैः[101] ॥ तत्र तावदेव अबजदचापं पूर्वोक्ता एव पूर्णज्याः संयोज्याः। अबपूर्णज्यार्धरूपं[102] अहमेकांशस्य जीवास्ति ॥ सा यावन्मिता कल्पिता या १ अबरेखा या २। एवं[103] बजं या २ जदं या २ यत एता मिथः समानाः सन्ति ॥ पुनः अदरेखा ६। १६। ४९। ७। ५९। ८। १६। अनया बजं या २ गुणितं अबजदघातयुतं जातं याव ४ या १२। ३३। ३८। १५। ५८। १६। ३३। ०। अयं जातः अजवर्गः ॥

पुनः प्रकारान्तरेणाजवर्गः[104] साध्यते। अबवर्गः याव ४। अयं व्या|[f. 5r J_2] सेन २ षष्टेरेकोर्ध्वपरिवर्तेन भक्तो लब्धं यावद्वर्गस्य कलाद्वयं २। इयमंशद्वयस्योत्क्रमज्या जाता।

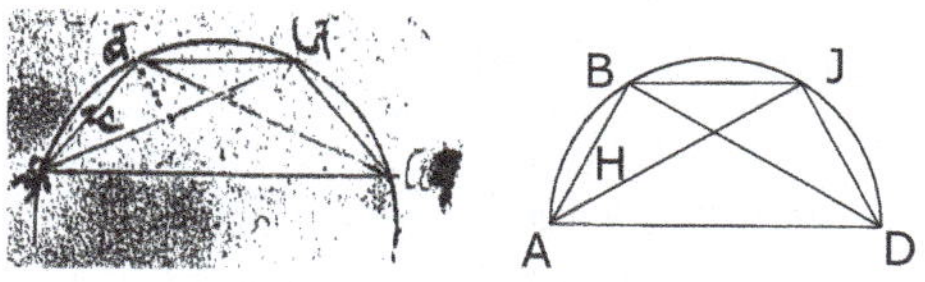

Figure J_2.4: Diagram on f. 4v of MS J_2 (facsimile and reconstruction)

Now here [is] the figure which [was] shown by Miryolugvega with the second method; in it the connection of many lines is made. Now here it is effected with only a few lines just as it is attained by the one called Mahamada Ābida. In that [method] just that much [is] arc $ABJD$; just the previously stated chords are to be connected. AH, having the form of half the chord of AB, is the sine of one degree. It is considered the amount of the *yāvattāvat*, *yā* 1, [and] line AB is *yā* 2. In the same way [line] BJ is *yā* 2 [and line] JD is *yā* 2, since they are equal to one another. Furthermore, line AD is 6; 16, 49, 7, 59, 8, 16. BJ [or] *yā* 2 is multiplied by that [and] added to the product of AB [and] JD; result, *yāva* 4, *yā* 12; 33, 38, 15, 58, 16, 33, 0. This result is the square of AJ.

Furthermore, the square of AJ is to be determined by another method. The square of AB is *yāva* 4. This is divided by the diameter, 2 once-raised by sixty; the quotient is two (2) minutes of the square of the *yāvattāvat*. This is the resulting versine of two degrees.

100 क्षेत्र J_2

101 ०आविदसंज्ञः J_2

102 ०रूपां J_2

103 एव J_2

104 ०वर्गः ॥ J_2

अस्या वर्गः यावव ४ विकला कबवर्गे शोधितः। शेषं याव ४ यावव ४̇[91] विकलाः। अयमंशद्वयज्यावर्गो जातः। अयं चतुर्गु|[f. 10r J_1] णितः याव १६ यावव १̇६ विक०। अयं सिद्धः कजवर्गः। एतौ पक्षौ समौ। समशोधनार्थं न्यासः।

याव ४ या १२। ३३। ३८। १५। ५८। १६। ३३

याव १६ यावव १̇६[92] विकलाः

शोधिते शेषम्।

याव १२

यावव १६ या १२। ३३। ३८। १५। ५८। १६। ३३

पुनर्द्वादशयावत्तावतापवर्त्य[93]।

या १

याघ[94] १ वि० २० रू १। २। ४८। ११। १९। ५१। २९। २५

इदं मिर्जोलुग्वेगेन निष्काशितं बहुप्रयासेन[95]।

Its square, *yāvava* 4 seconds, is subtracted from the square of *KB*. The remainder is *yāva* 4, *yāvava* −4 seconds. This is the resulting square of the sine of two degrees. This is multiplied by four: *yāva* 16, *yāvava* $-16''$. This is the determined square of *KJ*. These two sides [are] equal; for the purpose of subtraction of equals, the layout:

yāva 4 *yā* 12; 33, 38, 15, 58, 16, 33

yāva 16 *yāvava* −16 seconds

When subtracted, the remainder:

yāva 12

yāvava 16 *yā* 12; 33, 38, 15, 58, 16, 33

Again, reduced by twelve (12) times the *yāvattāvat*:

yā 1

yāgha $1''$, 20 *rū* 1, 2, 48, 11, 19, 51, 29, 25

This is expounded by Mirjolugvega with great effort.

91 ४ J_1
92 १६ J_1
93 पुनर्द्वादयाव० J_1
94 याव J_1
95 ०प्रयाशेन J_1

अस्या वर्गः यावव ४ विकला अबवर्गे शोधितः[105]। शेषं याव ४ यावव ४̇ विकलाः। अयमंशद्वयज्यावर्गो जातः ॥ अयं चतुर्गुणितः याव १६ यावव १̇६ वि०। अयं सिद्धः अजवर्गः ॥ एतौ पक्षौ समौ। समशोधनार्थं न्यासः।

याव ४ या १२। ३३। ३८। १५। ५८। १६। ३३[106]

याव १६ यावव १̇६ विकलाः

शोधिते शेषम्[107]।

याव १२

यावव १६ वि० या १२। ३३। ३८। १५। ५८। १६। ३३

पुनर्द्वादश १२ यावत्तावतापवर्तितः[108]।

या १

याघ १ वि २० प्रवि० रू १। २। ४८। ११। १९। ५१। २९। २५।

इदमेव मिर्योलुग्वेगेन निष्काशितं बहुप्रयासेन ॥

Its square, *yāvava* 4 seconds, is subtracted from the square of *AB*. The remainder is *yāva* 4, *yāvava* −4 seconds. This is the resulting square of the sine of two degrees. This is multiplied by four: *yāva* 16, *yāvava* $-16''$. This is the determined square of *AJ*. These two sides [are] equal; for the purpose of subtraction of equals, the layout:

yāva 4 *yā* 12; 33, 38, 15, 58, 16, 33

yāva 16 *yāvava* −16 seconds

When subtracted, the remainder:

yāva 12

yāvava 16 *yā* 12; 33, 38, 15, 58, 16, 33

Again, having reduced by twelve (12) times the *yāvattāvat*:

yā 1

yāgha $1'', 20'''$ *rū* 1, 2, 48, 11, 19, 51, 29, 25

This is expounded by Miryolugvega with great effort.

105 शोधितं J_2
106 ०३३। ० (and याव १६०—०विकलाः omitted) J_2
107 शोधिते शे J_2
108 पुनर्द्वादश १२ यावत्तावता ४ पवर्तितः J_2

2.8 [Dayānat Khān's Successive-Digits Iteration for the Sine of 1°]

।।[109] अथ दिआनतिखानस्य प्रकारो लिख्यते ।। तत्र समशोधने।
याव १२
यावव १६ वि० या १२। ३३। ३८। १५। ५८। १६। ३३

एतौ या ४ अपवर्तितौ[110]।
या ३
याघ ४ वि० रू ३। ८। २४। ३३। ५९। ३४। २८। १५

यद्यस्य रूपस्य[111] घनमूलं[112] ग्राह्यते परं च तादृशं घनमूलं ग्राह्यं[113] यस्याङ्कस्य वर्गो विकलाचतुष्टयेन गुणितो यावद्राशौ[114] शुद्धति। तच्छेषमुत्पादिताङ्केन[115] गुणितं रूपराशिसमं भवेत्[116]। एवेष्टाङ्को ऽस्ति ।।

Now the method of Diānatikhāna [Dayānat Khān] is written. In it, when subtraction of equals [is made]:
yāva 12
yāvava 16″ *yā* 12; 33, 38, 15, 58, 16, 33

These two reduced by *yā* 4:
yā 3
yāgha 4″ *rū* 3; 8, 24, 33, 59, 34, 28, 15

If the cube root of this constant is obtained (and indeed such a cube root is obtainable); [if] the square of whatever number multiplied by four seconds is subtracted from the coefficient of the *yāvattāvat*; [and if] the remainder from [subtracting] that, multiplied by the generated number, is equal to the amount of the constant; [then that] is just the desired number.

---

109 यद्धा या ६०
याघ ८ रू ५६
अत्र यावत्तावति ख २ in upper right margin with no evident insertion indicator, J_2

110 ४। पवर्त्तितौ J_2

111 घनो crossed out after रूपस्य J_2

112 कस्थ (?) crossed out after घनमूलं J_2

113 ८ (?) after ग्राह्यं J_2

114 यावद्रासौ J_2

115 तच्छेषं उत्पादि with उ inserted from right margin J_2

116 गुणितः रूपराशिसमो before भवेत् and शेषं केनाङ्केन गुणितः सन् रूपराशि तुल्यो भवति crossed out after it, J_2

अत्रोपपत्तिः ॥ इहैकपक्षे यावत्तावत्त्रयं द्वितीयपक्षे यावद्घन-विकलाश्चतस्रो रूपराशिश्च ॥ तत्र प्रथमपक्षे राशिद्वये योगो ऽस्त्यनुमीयते। यतो द्वितीयपक्षे राशिद्वयमस्त्यतो प्रथमपक्षे ऽपि राशिद्वयमपेक्षितमस्य समत्वात् ॥ तत्रैको राश्यङ्कस्तदन्तर्गत-स्तादृशो ऽस्ति य अपेक्षिताङ्कवर्गस्य विकलाचतुष्टयस्य च घा-ततुल्यो ऽस्ति। द्वितीयखण्डं तु तादृशमस्ति[117] येन प्रथमाङ्क-गुणितः सद्रूपराशितुल्यो[118] भवति ॥ ॥

अथ तदुत्पादनप्रकारः। तत्र रूपाणि स्थाप्यानि पुनर्यावनीय-सम्प्रदायेन चिह्नानि कृत्वा मूलं ग्राह्यं ॥ यथा प्रथमं रूपाणि

स्थापितानि। ३̇। ८। २४। ३̇३। ५९। ३४। २̇८। १५। पुनः खण्डद्वयं कोष्ठकानां कृतम्[119]। अत्रैकाङ्कघन १ उत्पादितः स च तृतीयपङ्क्तौ स्थापितः। अस्य वर्गः १| [f. 5v J_2] विक-लाचतुष्टयेन गुणितः ४ द्वितीयपङ्क्त्यधो[120] लिखितः ॥ अयं यावत्तावतांशत्रये या ३ शोधितः शेषं २। ५९। ५६।

Here is the demonstration. Here in one side [are] three *yāvat-tāvats*; in the second side four seconds of the cube of the *yāvattāvat* and the amount of the constant. There in the first side is a sum in [the combination of] a pair of quantities: [thus it] is inferred. Since in the second side there is a pair of quantities, hence in the first side too a pair of quantities is presumed, because of its equality [with the second side]. There, one quantity-number included in it is as much as is equal to the product of the square of the produced number and four seconds. But the second part is as much as with which, being multiplied by the first number, is equal to the amount of the constant.

Now the method for generating that. Having made the constant [digits] placed there [and] also the dots by the Yāvanī established usage, the root is to be obtained. As first, the constant [digits] are placed:

$\dot{3}; 8, 24, \dot{3}3, 59, 34, \dot{2}8, 15.$

Furthermore, two [more] sections of grid cells are made. Here the cube of one digit [or, the number one], 1, is generated, and it is set in the third *paṅkti*. Its square, 1, multiplied by four seconds, 4, is written below the second *paṅkti*. This is subtracted from three degrees of the *yāvattāvat* (*yā* 3), remainder: 2; 59, 56.

[117] तु inserted before तु (?) from bottom margin, तादृशास्ति with म inserted from bottom margin J_2

[118] सन् रूप० J_2

[119] कोष्टाकानकृतं J_2

[120] द्वितीयपङ्क्धो J_2

पुनर्लब्ध्या[121] गुणितो रूपेषु शोधनार्थं प्रथमपङ्क्तिसमो[122] लिखितः ॥ २। ५९। ५६। शोधिते ऽवषिष्टं ८। २८। शेषं पूर्वस्थान्येव ॥ अथ लब्धिः १ तृतीयपङ्क्तिस्थलब्धौ १ योजितः २ तृतीयपङ्क्तिसमो लिखितः ॥ पुनरयं २ प्रथमलब्ध्या[123] १ गुणितः २ विकलाभिः ४ गुणितश्च वि ८।

अयं द्वितीयपङ्क्तिस्थविकलास्थानसमो न्यस्तः ८। शोधिते ऽवशिष्टमेककोष्ठकमुत्तार्य[124] लिखितं २। ५९। ४८। पुनः प्रथमलब्धिः १ निजाधःस्थाङ्के २[125] युता ३। इयमेककोष्ठकमुत्तार्य[126] लिखिता[127] ३।

पुनः प्रथमपङ्क्तौ द्वितीयचिह्ने ३३ एकाङ्क उत्पादितः। स च तस्योपरिस्थः ॥ तत्सन्मुखे तृतीयपङ्क्तौ च २। पुनर्लब्धिः २ तृतीयपङ्क्तिस्थिताङ्काभ्यां ३। २ गुणिता ६। ४। विकलाभि ४ र्गुणिताश्च २४। १६। एताः कला २ नां धातत्वात्प्रतिविकलाः प्र॰वि॰[128] २४। १६ जाताः।

[That] is further multiplied by the quotient [and] for the purpose of subtraction from the constant [digits] written on the level of the first *paṅkti*: 2; 59, 56. When subtracted, remainder: 8, 28. The remainder is just [the digits] standing in front [in the constant]. Now the quotient 1 added to the quotient in the third *paṅkti*, 1, [making] 2, is written on the level of the third *paṅkti*. Again, this 2, multiplied by the first quotient 1, [making] 2, and multiplied by 4 seconds, is 8″.

This is set down on the level of the seconds place standing in the second *paṅkti*: 8. When subtracted, the remainder is written passing over one grid cell: 2, 59, 48. Again, the first quotient 1 is added to the number standing below itself, 2, [making] 3. This is written passing over one grid cell: 3.

Again, at the second dot [over] 33 in the first *paṅkti*, one digit is generated. And it is standing above that; and opposite it in the third *paṅkti*, 2. Again, the quotient 2 is multiplied by the two numbers 3, 2 standing in the third *paṅkti*: 6, 4. And multiplied by 4 seconds, [they are] 24, 16. Because of [their] being a product with 2 minutes, these are thirds: 24‴, 16.

---

121 पुन लब्ध्या J_2

122 प्रथमपङ्क्ति॰ J_2

123 ल inserted from left margin J_2

124 मुर्त्तार्य J_2

125 निजाधस्थां २ के J_2

126 अयमे॰ with अ crossed out, इ inserted from left margin J_2

127 लिखितः J_2

128 घातत्वा then प्र॰ above line, then विकलाः then प्र वि above line, J_2

द्वितीयपङ्क्त्यां शोधनार्थं निजस्थाने स्थापिताः ॥ शोधिते ऽवशिष्टं २। ५९। ४७। ३५। ४४। पुनः शेषमिष्टलब्ध्या[129] २ कलागुणितं ५। ५९। ३५। ११। २८ प्रथमपङ्क्त्यधो यथा स्थाने लिखितं शोधिते ऽवशिष्टं २। २८। ५८। ४८। ६। पुनर्लब्धिरधःस्थिताङ्के २ युक्ता ४। इयं तदधो लिखिता ४। पुनः पूर्वलब्धिः २ अनेनाङ्केन ३। ४ गुणिता विकलाभिश्च ४ गुणिता २४। ३२।

एताः प्रतिविकलाः स्वस्थाने द्वितीयपङ्क्त्यां[130] स्थापिताः। शोधिते शिष्टं २। ५९। ४७। ११। १२। इदमेककोष्ठकमुत्तार्य द्वितीयपङ्क्तौ लिखितम् ॥ पुनर्लब्धिः २ तृतीयपङ्क्तिस्थकलासु ४ योजिताः ३। ६। इदमेककोष्ठमुत्तार्य तदधः ३। २ लिखितम् ॥

पुनस्तृतीयचिह्ने २८ इष्टाङ्क उत्पादितः। स च ४९ तदुपरिस्थस्तृतीयपङ्क्तौ तदधःस्थश्च। एवमधो ऽङ्कः ३। ६। ४९ । जातः पुनर्लब्ध्या गुणितः[131] विकलाभि ४ र्गुणितश्च १०। १०। १६। ४। प्रतिविकला एताः[132] जाताः ॥

For the purpose of subtraction they are set in the second *paṅkti* in their own place. When subtracted, remainder: 2, 59, 47, 35, 44. Again, the remainder multiplied by the desired quotient 2 minutes, [making] 5, 59, 35, 11, 28, is written below the first *paṅkti* according to place; when subtracted, remainder: 2, 28, 58, 48, 6. Again, the quotient is added to the number standing below, 2, [making] 4. This is written below that: 4. Again, the previous quotient 2 is multiplied by that number 3, 4 and multiplied by 4 seconds: 24, 32.

These thirds are set in their own place in the second *paṅkti*. When subtracted, remainder: 2, 59, 47, 11, 12. This is written passing over one grid cell in the second *paṅkti*. Again, the quotient 2 is added to the 4 minutes standing in the third *paṅkti*: 3, 6. This is written below that, passing over one grid cell, 3, 2.

Further, at the third dot [over] 28, the desired digit is generated. And that, 49, is standing above it and standing below it in the third *paṅkti*. In this way the number below, 3, 6, 49, results, and again is multiplied by the quotient and multiplied by 4 seconds: 10, 10, 16, 4. These are the resulting thirds.

---

129 ष्ट inserted from left margin J_2
130 द्विप० पङ्क्ता J_2
131 जाता पुनर्लब्ध्या गुणिताः J_2
132 प्रतिविकलैताः J_2

इदं द्वितीयपङ्क्तौ शोधनार्थं न्यस्तम् ॥ शोधिते शेषं २। ५९। ४७। १। १। ४३। ५६। पुनरिदं लब्ध्या ४९। गुणितं २। २६। ४९। २३। ५०। २४। ५२। ४४। रूपेषु शोधिते शिष्टं ॥ २। ९। २४। १६। ३। २२। १६। अथ लब्धिः ४९ निजाधःस्थतृतीयपङ्क्तिलब्धौ ४९ योजिता। जातः ३। ७। ३८। पुनरिदं लब्ध्या गुणितं विकला ४ भिश्च गुणितं १०। १२। ५६[133]। ८।

इदं द्वितीयपङ्क्त्यां न्यस्तम् ॥|[f. 6r J_2] पुनस्तत्र शोधिते शिष्टमेककोष्ठकमुत्सार्य लिखितं २। ५९। ४६। ५०। ४८। ४७। ४८। अत्र लब्धिः ४९ तृतीयपङ्क्तिस्थाङ्के ३८ योजिता ३। ८। २७ एककोष्ठमुत्तार्य लिखिता ॥

पुनश्चतुर्थाङ्के चिह्ने ० अङ्क उत्पादितः ४३ तदुपरि लिखितस्तृतीयपङ्क्त्यधश्च। एवमधःस्याङ्कः ३। ८। २७। ४३ जातः। अयं लब्ध्या ४३ गुणितो विकलाभि ४ र्गुणितश्च ९। ०। १६ द्वितीयपङ्क्त्यां शोधितः २। ५९। ४६। ५०। ३९। ४७। ३२। पुनरिदं लब्ध्या ४३ गुणितं २। ८। ५०। ३४। १८। ३१। ४। १७। प्रथमपङ्क्तिस्थरूपेषु शोधितः शे० ३३। ४१। ४४। ५१। ११। ४३।

For the purpose of subtraction this is set in the second *paṅkti*. When subtracted, remainder: 2, 59, 47, 1, 1, 43, 56. Again, this is multiplied by the quotient 49: 2, 26, 49, 23, 50, 24, 52, 44. When subtracted from the constant [digits], remainder: 2, 9, 24, 16, 3, 22, 16. Now the quotient 49 is added to the quotient 49 in the third *paṅkti* standing below itself; result, 3, 7, 38. Again, this is multiplied by the quotient and multiplied by 4 seconds: 10, 12, 56, 8.

This is set in the second *paṅkti*. Again, when subtracted there, the remainder is written passing over one grid cell: 2, 59, 46, 50, 48, 47, 48. Here the quotient 49 is added to the number 38 standing in the third *paṅkti*: 3, 8, 27 is written passing over one grid cell.

Furthermore, at the dot [over] 0 at the fourth [dotted] digit, the generated digit 43 is written above that and below [in] the third *paṅkti*. In this way the number standing below, 3, 8, 27, 43, results. This multiplied by the quotient 43 and multiplied by 4 seconds, [making] 9, 0, 16, is subtracted in the second *paṅkti*: 2, 59, 46, 50, 39, 47, 32. Again, this is multiplied by the quotient 43: 2, 8, 50, 34, 18, 31, 4, 17. Subtracted from the constant [digits] standing in the first *paṅkti*: rem., 33, 41, 44, 51, 11, 43.

---

[133] १६ J_2

पुनर्ल० ४३ तृतीयपङ्क्त्यधःस्थलब्धौ योजिता ३ । ८ । २८ । २६।
पुनरिदं लब्ध्या[134] गुणितं विकलाभिश्च गुणितं ९ । ० । १७। इदं
शोधितं शेषं २ । ५९ । ४६ । ५० । ३० । ४७। इदमेककोष्ठमुत्तार्य
लिखितम् ॥ पुनर्लब्धि ४३ तृतीयपङ्क्त्यधःस्थाङ्के २६ युक्ता जा०
३ । ८ । २९ । ९ । अयमेककोष्ठमुत्तार्य लिखितः ॥

पुनः पञ्चमचिह्ने ० इष्टाङ्क उत्पादितः ११। अयं स्वोपरि तथा
तृतीयपङ्क्तौ च लिखितः ३ । ८ । २९ । ९ । ११ लब्ध्या ११ गु-
णितो विकलाभि ४ र्गुणितश्च २[135] । शोधिते शिष्टं २ । ५९ ।
४६ । ५० । ३० । ४५। पुनर्लब्ध्या ११ गुणितः ३२ । ५७ । ३५ ।
१५ । ३८ । १५। रूपेषु शोधितं शे० ४४ । ९ । ३५ । ३३ । ४९।
पुनर्लब्धिस्तृतीयस्थ ११ योजिता ३ । ८ । २९ । ९ । २२। लब्ध्या
११ गुणितो विक०४ गुणितश्च २। शोधिते शिष्टमेककोष्ठमुत्तार्य
लिखितं २ । ५९ । ४६ । ५० । ३१।

अयमर्धाभ्यधिकः। पुनर्लब्धिस्तृतीयस्थ २२ योजिता ३ । ८ ।
२९ । ९ । ३३। अयमप्येकाङ्कमुत्तार्य लिखितः ॥

Again, quo. 43 is added to the quotient standing below [in] the third *paṅkti*: 3, 8, 28, 26. Again, this is multiplied by the quotient and multiplied by the seconds: 9, 0, 17. This is subtracted: remainder, 2, 59, 46, 50, 30, 47. This is written passing over one grid cell. Again, quotient 43 is added to the number standing below [in] the third *paṅkti*, 26: res., 3, 8, 29, 9. This is written passing over one grid cell.

Furthermore, at the fifth dot [over] 0 the desired digit is produced, 11. This is written above its own [place] and likewise in the third *paṅkti*, 3, 8, 29, 9, 11, multiplied by the quotient 11 and multiplied by 4 seconds, 2. When subtracted, remainder: 2, 59, 46, 50, 30, 45. Again, multiplied by the quotient 11: 32, 57, 35, 15, 38, 15. Subtracted from the constant [digits], rem.: 44, 9, 35, 33, 49. Again, the quotient added to 11 standing in the third [*paṅkti*]: 3, 8, 29, 9, 22. Multiplied by the quotient 11 and multiplied by 4″: 2. When subtracted, the remainder is written passing over one grid cell: 2, 59, 46, 50, 31.

This is greater than a half. Again, the quotient is added to 22 standing in the third [*paṅkti*]: 3, 8, 29, 9, 33. This too is written passing over one digit.

[134] पुन inserted from right margin, then इदं लब्ध्यां J_2
[135] विकला ४ र्गुणितश्च २ J_2

पुनः षष्टचिह्ने ० इष्टाङ्क[136] उत्पादितः १४ तदुपरि तथा तृतीयपङ्क्तौ च लिखि० ३ । ८ । २९ । ९ । ३३ । १४। अयं लब्ध्या १४ गुणितो विकलाभिश्च ० । ०[137]। शोधितः शे० २ । ५९ । ४६ । ५० । ३१ । लब्ध्या गुणितः ४१ । ५६ । ५५ । ४७ । १४। इदं रूपे शो० २ । १२ । ३९ । ४६ । ३५ । एवमग्रे ऽपि । | [f. 6v J_2]

Furthermore, at the sixth dot [over] 0 the desired digit is generated, 14, writ. above that and likewise in the third *paṅkti*: 3, 8, 29, 9, 33, 14. This is multiplied by the quotient 14 and by the seconds: 0, 0. [When] subtracted, rem.: 2, 59, 46, 50, 31. Multiplied by the quotient: 41, 56, 55, 47, 14. This sub. from the constant: 2, 12, 39, 46, 35. And in the same way subsequently.

---

[136] इव्यंक J_2

[137] ० । ० followed by तत्स्थानपर्यतमंकोनसमागमः अतो शून्यं struck out J_2

	१			२			४९			४३			११			१४
प्र	३	८	२४	३३	५९	३४	२८	१५	०	०	०	०	०	०	०	०
	२	५९	५६													
		८	२८													
		५	५९	३५	११	२८	[138]									
		२	२८	५८	४८	६										
		२	२६	४९	२३	५०	२४	५२	४४							
			२	९	२४	१६	३	२२	१६	[139]						
			२	८	५०	३४	१८	३१	४	१७						
					३३	४१	४४	५१	११	४३						
					३२	५७	३५	१५	३८	१५						
						४४	९	३५	३३	४९	[140]					
						४१	५६	५५	४७	१४						
						२	१२	३९	४६	३५						
द्वि	३	०	०													
			४													
	२	५९	५६													
			८	[141]												
		२	५९	४८												
					२४	१६										
		२	५९	४७	३५	४४										
					२४	३२										
			२	५९	४७	११	१२	[142]								
						१०	१०	१६	४							
			२	५९	४७	१	१	४३	५६							
						१०	१२	५६	८	[143]						
				२	५९	४६	५०	४८	४७	४८	[144]					
								९	०	१६	[145]					
				२	५९	४६	५०	३९	४७	३२						
								९	०	१७						
					२	५९	४६	५०	३०	४७						
					२	५९	४६	५०	३०	४५	[146]					
						२	५९	४६	५०	३१						
तृ	१		३	२	३	६	४९									
	२			४	३	७	३८									
							३	८	२७	४३						
									२८	२६						
									३	८	२९	९	११			
													२२			
											३	८	२९	९	३३	१४

Table 2.1: See figure 3.8 for this calculation grid as it appears in MS J_2 and figure 3.7 for its translation. See section 3.8 for an analysis of the layout and numerical entries.

138 ५ in leftmost column written over ७ J_2

139 २ crossed out in column before leftmost J_2

140 First three digits originally written one column to the left and corrected J_2

141 ८ squeezed in between rows J_2

142 १२ omitted J_2

143 १६ written for ५६ J_2

144 First five digits originally written one column to the left and corrected J_2

145 Three digits originally written one column to the left and corrected J_2

146 ४२ written for ४५ J_2

Chapter 3

Mathematical Notes on the *jyācāpa* Excerpt

The iterative approximation techniques described in the Jaipur manuscript's excerpt of the *jyācāpa* section of the *Siddhānta-kaustubha* can be used for determining either $R \sin 1°$, denoted x, or its double Crd $2°$, denoted y. As explained in section 1.1, $R \sin$ denotes the standard sine function scaled to the given trigonometric radius R (usually 60 in this context), and Crd is the chord of the designated arc in the circle of radius R. In more general versions of these techniques applied to computing the sine or double-sine of an arc $\alpha/3$ where α is an arbitrary arc whose $R \sin$ is known, we denote the desired quantities by $X = R \sin(\alpha/3)$ and $Y = 2R \sin(\alpha/3)$.

Section	Ch. 2 section	Iteration method	Quantity/ies computed	Sources and connections
3.1	2.1	Successive digits	$Y = 2R \sin(\alpha/3)$	*Siddhāntakaustubha* 14–17/II.1.62–65 (§ 5.2.2.1)
3.2	2.2	Successive values	$Y = 2R \sin(\alpha/3)$, $X = R \sin(\alpha/3)$	*Siddhāntakaustubha* 18–19/II.1.66–67 (§ 5.2.2.2)
3.3	2.3	Successive digits	$y = \text{Crd } 2°$ ($R = 60$)	Attributed to al-Kāshī, via al-Bīrjandī (§ 4.3.3)
3.4	2.4	Successive values	$y = \text{Crd } 2°$ ($R = 60$)	Attributed to Muḥammad Ābida
3.5	2.5	Successive values	$y = \text{Crd } 2°$ ($R = 60$)	Attributed to Ulugh Beg, via al-Bīrjandī (§ 4.3.4)
3.6	2.6	Successive values	$x = R \sin 1°$ ($R = 60$)	Attributed to Ulugh Beg, via al-Bīrjandī (§ 4.3.4)
3.7	2.7	Alternate proof of previous successive-values iteration for $x = R \sin 1°$		Attributed to Muḥammad Ābida
3.8	2.8	Successive digits	$x = R \sin 1°$ ($R = 60$)	Attributed to Dayānat Khān

Table 3.1: Iterative techniques described in the *jyācāpa* excerpt.

The author(s) expound seven methods for computing these quantities in varying levels of detail (along with an additional proof for one of them). Some involve what we have called in section 1.1.2 "successive-digits" algorithms, while the rest are of the "successive-values" type. The subsequent sections of this chapter describe each of them in order. Table 3.1 summarizes their essential content and interrelationships, along with other information about related texts.

3.1 Jagannātha's Successive-Digits Iteration for Y, §2.1

3.1.1 Methods

Preserved only in MS J_1, this portion closely glosses the description of the successive-digits iteration technique for $Y = 2R \sin(\alpha/3)$ stated by Jagannātha in *Siddhāntakaustubha jyācāpa* verses *SK* II.1.62–65/J_S 14–17 (see section 5.2). The implied cubic equation is

C. Montelle et al., *From Samarqand to Jaipur*, Sources and Studies in the History of Mathematics and Physical Sciences, https://doi.org/10.1007/978-3-031-89087-1_3

$$Y = \frac{2R\sin\alpha}{3} + \frac{Y^3}{3R^2}. \tag{3.1}$$

This generalization of the standard cubic equation for y introduced in equation 1.4 replaces 3° by an arbitrary arc α with known $R\sin$, and substitutes for the standard Greco-Islamic radius $1,0 = 60$ an arbitrary trigonometric radius R.

The procedure prescribes a cycle of steps for deriving the successive sexagesimal digits $\{q_0, q_1, q_2, \ldots\}$ of the approximate value $Y \approx q_0; q_1, q_2, \ldots$ in the *paṅkti* or "output row". The crux of this approximation technique, as of all the other successive-digits approximations descended from al-Kāshī's method, is the substitution of known quantities for the unknown Y^3 term in the original equation for Y. Effectively, it involves breaking down Y^3 into the sum

$$Y^3 = (q_0; q_1, q_2, \ldots)^3 + (Y^3 - (q_0; q_1, q_2, \ldots)^3), \tag{3.2}$$

using successively improved approximate values for Y in the difference-of-cubes term.

The inaccuracy in this difference-of-cubes term representing $(Y^3 - (q_0; q_1, q_2, \ldots)^3)$ continues to shrink as more sexagesimal digits are added to the approximate value of Y, and eventually becomes small enough to be able to be neglected. Our reconstruction of this iteration version's rationale is laid out together with sample calculations in the table in the following section.

3.1.2 Calculations

No numerical computations are included in the author's account. To illustrate the prescribed procedure, and for ease of comparison with our reconstruction of a similar procedure in section 3.3.2, we have carried out two iterations of it on the parameter values

$$R = 60, \quad \alpha = 3°, \quad \text{and } R\sin 3° = 3; 8, 24, 33, 59, 34, 28, 14, 50,$$

laid out with step-by-step explanations in the following table. (Here and throughout our discussion we use the division-sign obelus "÷" to indicate integer division, i.e., the floor function of ordinary division. A remainder accompanying the integer quotient is prefixed by "*r*".)

"Having determined the sine of the arc... multiplied by two, and divided by three... set those down separately (the result is degrees)..." Current *paṅkti*: $q_0 = 2$	$R\sin 3^\circ = 3;8,24,33,59,34,28,14,50$ $(2\cdot R\sin 3^\circ)\div 3$ $= (6;16,49,7,59,8,56,29,40)\div 3$ $= 2\quad r\ 0;16,49,7,59,8,56,29,40$
"...the cube of those... degrees... is divided by the square of the half-diameter .. increased by the remainder from [dropping] the degrees..."	$\frac{2^3}{(60)^2} = 0;0,8$ $\begin{array}{l} 0;16,49,7,59,8,56,29,40\\ +\ 0;\ 0,\ 8\\ \hline 0;16,57,7,59,8,56,29,40\end{array}$
"divided by three... the result in minutes, that is to be set down below the result... that is the *paṅkti* produced in degrees and minutes..." Current *paṅkti*: $q_0;q_1 = 2;5$	$(0;16,57,7,59,8,56,29,40)\div 3$ $= 0;5\quad r\ 0;1,57,7,59,8,56,29,40$
"The cube of the *paṅkti* diminished by the first cube is divided by the square of the half-diameter... added to the remainder..."	$\frac{(2;5)^3-2^3}{(60)^2} = 0;0,1,2,32,5$ $\begin{array}{l} 0;1,57,\ 7,59,\ 8,56,29,40\\ +\ 0;0,\ 1,\ 2,32,\ 5\\ \hline 0;1,58,10,31,13,56,29,40\end{array}$
"...by three... the resulting quotient in seconds is to be set below the minutes... That resulting *paṅkti* [has] the form degrees, minutes, seconds..." Current *paṅkti*: $q_0;q_1,q_2 = 2;5,39$	$(0;1,58,10,31,13,56,29,40)\div 3$ $= 0;0,39\quad r\ 0;0,1,10,31,13,56,29,40$
"... The cube of the *paṅkti* is to be made. The cube of the previous [*paṅkti*] is to be subtracted from that... The previously-stated [procedure] is to be done... again and again..." Current *paṅkti* $q_0;q_1,q_2,q_3 = 2;5,39,26$	$\frac{(2;5,39)^3-(2;5)^3}{(60)^2} = 0;0,0,8,30,27,27,43,39$ $\begin{array}{l} 0;0,1,10,31,13,56,29,40\\ +\ 0;0,0,\ 8,30,27,27,43,39\\ \hline 0;0,1,19,\ 1,41,24,13,19\end{array}$ $(0;0,1,19,1,41,24,13,19)\div 3$ $= 0;0,0,26\quad r\ 0;0,0,1,1,41,24,13,19$

We compute the first approximation $Y \approx \frac{2R\sin\alpha}{3}$. Representing the quotient as the sum of the first *paṅkti* digit q_0 and one-third of the remainder from the integer division, $\frac{1}{3}r_0$:	$q_0 = 2R\sin\alpha \div 3$ $Y \approx q_0 + \frac{1}{3}(2R\sin\alpha - 3q_0)$ $= q_0 + \frac{1}{3}r_0$
We can view this step as improving the approximation by including a version of the original equation's cubic term, substituting ${q_0}^3$ for Y^3:	$Y = q_0 + \frac{1}{3}r_0 + \frac{Y^3}{3R^2}$ $= q_0 + \frac{1}{3}r_0 + \frac{{q_0}^3}{3R^2} + \frac{Y^3 - {q_0}^3}{3R^2}$ $\approx q_0 + \frac{1}{3}\left(r_0 + \frac{{q_0}^3}{R^2}\right) + 0$
The integer division by 3 now produces the next *paṅkti* digit q_1, with its corresponding remainder-third $\frac{1}{3}r_1$:	$0;q_1 = \left(r_0 + \frac{{q_0}^3}{R^2}\right) \div 3$ $Y \approx q_0;q_1 + \frac{1}{3}\left(r_0 + \frac{{q_0}^3}{R^2} - (3)(0;q_1)\right)$ $= q_0;q_1 + \frac{1}{3}r_1$
We can now incorporate a new version of the difference-of-cubes term $\frac{Y^3 - {q_0}^3}{3R^2}$ by using the new value $q_0;q_1$ in place of Y, and factor out the divisor 3 again:	$Y = q_0 + \frac{1}{3}\left(r_0 + \frac{{q_0}^3}{R^2}\right) + \frac{Y^3 - {q_0}^3}{3R^2}$ $= q_0;q_1 + \frac{1}{3}r_1 + \frac{Y^3 - {q_0}^3}{3R^2}$ $\approx q_0;q_1 + \frac{1}{3}r_1 + \frac{(q0;q1)^3 - {q_0}^3}{3R^2}$ $= q_0;q_1 + \frac{1}{3}\left(r_1 + \frac{(q0;q1)^3 - {q_0}^3}{R^2}\right)$
Again, integer division by 3 produces the next *paṅkti* digit q_2 and adjusted remainder term $\frac{1}{3}r_2$. The resulting approximation is tweaked with a sharpened-up value of the difference-of-cubes term using the cubed *paṅkti* values from the two most recent iterations. The same cycle of steps can be repeated for as many digits of Y as desired.	$0;0,q_2 = \left(r_1 + \frac{(q0;q1)^3 - {q_0}^3}{R^2}\right) \div 3$ $Y \approx q_0;q_1,q_2$ $+ \frac{1}{3}\left(r_1 + \frac{(q0;q1)^3 - {q_0}^3}{R^2} - (3)(0;0,q_2)\right)$ $= q_0;q_1,q_2 + \frac{1}{3}r_2$ $Y \approx q_0;q_1,q_2 + \frac{1}{3}\left(r_2 + \frac{(q0;q1,q_2)^3 - (q_0;q_1)^3}{R^2}\right)$ ...

3.2 Jagannātha's Successive-Values Iterations for Y and X, §2.2

3.2.1 Methods

Likewise found only in MS J_1, these two successive-values iteration algorithms mirror the ones for $Y = 2R\sin(\alpha/3)$ and for $X = R\sin(\alpha/3)$ in *Siddhāntakaustubha* verses *SK* II.1.66 and 67/J_S 18 and 19 respectively (see section 5.2). The first method (J_S verse 18) begins by setting $Y_0 = \frac{2}{3}R\sin\alpha$ ("the sine of the arc diminished by its own third part"), which is equal to $\text{Crd}(2\alpha)/3$. Then we apply the following iterative equation, which is equivalent to generalizing its counterpart in equation 1.5 to arbitrary R and α:

$$Y_{k+1} = Y_0 + \frac{(Y_k)^3}{3R^2}, \quad k \in \{0, 1, 2, 3, \ldots\}. \tag{3.3}$$

The *paṅkti* Y is equal to $\text{Crd}(2\alpha/3)$, so half of the *paṅkti* is the desired $R\sin(\alpha/3)$.

The second method (J_S verse 19) sets $X_0 = \frac{1}{3}R\sin\alpha$, and iterates according to the equation

$$X_{k+1} = X_0 + \frac{(X_k)^3 \cdot 4/3}{R^2}, \quad k \in \{0, 1, 2, 3, \ldots\}, \tag{3.4}$$

similarly generalizing its counterpart in equation 1.5.

3.2.2 Calculations

As in the preceding section, no numerical computations are included in the author's account of Jagannātha's successive-values methods. We illustrate in the following table three iterations of the author's procedure for Y using the same sample parameters as before: $R = 60$, $\alpha = 3°$, and $R\sin 3° = 3; 8, 24, 33, 59, 34, 28, 14, 50$. In contrast to the previous successive-digits method, the *paṅkti* is entirely updated in the successive-values method after each iteration.

Since Jagannātha's successive-values method for X is almost identical to the one for Y, we have not illustrated it separately with a sample numerical calculation.

"The sine of the given arc, diminished by its own third part, is to be set down separately. . ."	$\frac{2 \cdot R \sin 3^\circ}{3} = 2;5,36,22,39,42,58,49,53,20$
"The cube of it is divided by three times the square of the radius. . ."	$\frac{(2;5,36,22,39,42,58,49,53,20)^3}{3(60)^2} = \frac{9;10,28,3,8,52,5,36,54,8,44,22,20}{3(60)^2} = 0;0,3,3,29,21,2,57,21,52$
"The diminished sine of the arc standing separately is to be increased by the resulting quotient. . ." Current *paṅkti*: 2;5,39,26,9,4,1,47,15,12	$\begin{array}{r} 2;5,36,22,39,42,58,49,53,20 \\ +\ 0;0,\ 3,\ 3,29,21,\ 2,57,21,52 \\ \hline 2;5,39,26,\ 9,\ 4,\ 1,47,15,12 \end{array}$
"Repeatedly with that procedure a *paṅkti* is to be determined. . ."	$\frac{(2;5,39,26,9,4,1,47,15,12)^3}{3(60)^2} = \frac{9;11,8,16,32,30,47,52,32,14}{3(60)^2} = 0;0,3,3,42,45,30,50,15,57$
Current *paṅkti*: 2;5,39,26,22,28,39,40,9,17	$\begin{array}{r} 2;5,36,22,39,42,58,49,53,20 \\ +\ 0;0,\ 3,\ 3,42,45,30,50,15,57 \\ \hline 2;5,39,26,22,28,39,40,\ 9,17 \end{array}$
	$\frac{(2;5,39,26,22,28,39,40,9,17)^3}{3(60)^2} = \frac{9;11,8,19,28,58,14,30,46,24}{3(60)^2} = 0;0,3,3,42,46,29,39,24,50$
Current *paṅkti*: 2;5,39,26,22,29,28,29,18,10	$\begin{array}{r} 2;5,36,22,39,42,58,49,53,20 \\ +\ 0;0,\ 3,\ 3,42,46,29,39,24,50 \\ \hline 2;5,39,26,22,29,28,29,18,10 \end{array}$

3.3 Al-Kāshī's Successive-Digits Iteration for y, §2.3

3.3.1 Methods

This approximation approach is ascribed in J_2 to al-Kāshī himself via al-Bīrjandī's commentary, which is discussed in its original Persian form in section 4.3.1. The procedure is described almost identically in J_1 and J_2, except for J_2's omission of a part of its geometric demonstrations. Therefore, in our analysis of this section we follow J_1's more "Sanskritized" convention of using the letter label K (Sanskrit *ka*) instead of A in the diagrams. The exposition of the method combines the following fundamental components:

Known values	of $R \sin 3°$ and hence its double Crd 6°, scaled with high precision to the trigonometric radius $R = 60$ and represented as line segments in a cyclic quadrilateral diagram.
Two geometric premises	involving cyclic quadrilaterals, which determine a theoretical relation between powers of the desired unknown quantity and the known trigonometric function values: 1. "Ptolemy's theorem" on diagonals and sides of cyclic quadrilaterals, the second diagram in the first chapter of the *Almagest* (Toomer 1984, I.2, p. 50). Our text refers to the *Almagest* as *Mijistī*, or else by the name of Jagannātha Samrāṭ's Sanskrit translation of it, *Siddhāntasamrāṭ*. 2. The "half-arc theorem" of the fourth diagram in the first chapter of the *Almagest* (Toomer 1984, I.4, p. 52).
Algebraic operations	for simplifying the above relation to a cubic equation containing the unknown quantity, its cube, and a constant term.
An iterative algorithm	for approximating the solution to the cubic equation by generating successive sexagesimal digits added to the originally assumed estimate of the unknown.

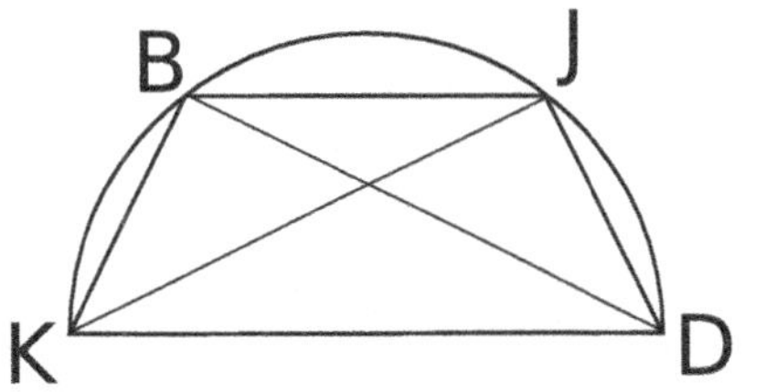

Figure 3.1: Diagram reproduced from figure J_1.2

Known Values Our text begins its explanation by describing the cyclic quadrilateral labeled $KBJD$ in MS J_1 and pictured in figure 3.1. Its base KD is Crd 6°, while each of its other three sides KB, BJ, JD equals the unknown Crd 2°. The known trigonometric quantities are declared to be

$$R \sin 3° = 3; 8, 24, 33, 59, 34, 28, 14, 50 \qquad \text{and} \qquad KD = \text{Crd}\, 6° = 2\,R \sin 3° = 6; 16, 49, 7, 59, 8, 56, 29, 40.$$

Two Geometric Premises The text's "first premise", namely Ptolemy's theorem, states that the sum of the products of each pair of opposite sides in the quadrilateral is equal to the product of its diagonals KJ and BD:

$$KB \cdot JD + BJ \cdot KD = KJ \cdot BD. \tag{3.5}$$

Version J_1 claims that the rationale or demonstration of this theorem was "just explained previously", although no such explanation survives in our manuscript. Like the demonstration of the "second premise" described below, it probably derives ultimately from an earlier portion of the Bīrjandī commentary.

The "second premise" relates the chord of the supplement of an arc α to the chord of $\alpha/2$ in a circle of radius R as follows:

$$(\text{Crd}\,\alpha/2)^2 = R \cdot (2R - \text{Crd}(180° - \alpha)). \tag{3.6}$$

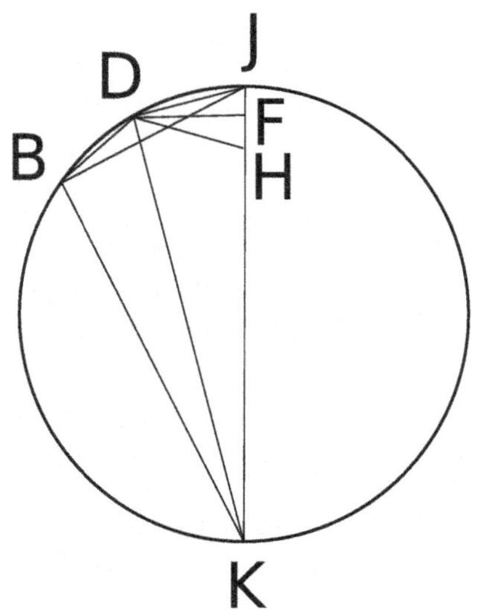

Figure 3.2: Diagram reproduced from figure $J_1.1$

The following demonstration, which occurs only in MS J_1, invokes the diagram in figure 3.2. There the given arc α is $\widehat{BJ}$ halved at D, so $\widehat{BD} = \widehat{DJ}$; and $KJ = 2R$, implying that $\widehat{KB} = 180° - \alpha$. Drawing $DF \perp KJ$, we claim that $JF = \frac{1}{2}(KJ - KB)$. The argument for this claim is as follows:

Let $KH = KB$, and join DH. Then triangles HKD and DKB are equal, because $KH = KB$, side KD is shared, and $\angle HKD = \angle BKD$. Therefore $DH = BD = DJ$.

Because triangles DJF and DHF are right, $DJ^2 = DF^2 + JF^2$ and $DH^2 = DF^2 + FH^2$, so $JF = FH$. Their sum $FH + JF = 2JF$ is $JH = KJ - KH$. Since $KH = KB$, JF is indeed one half of $KJ - KB$.

Moreover, the angle D in triangle KDJ on the diameter KJ is right (although the author does not explicitly invoke Thales' theorem to support this claim). Since triangles KDJ and JDF are similar, $KJ/JD = JD/JF$. So $JD^2 = KJ \cdot JF$, or

$$(\mathrm{Crd}\,\alpha/2)^2 = 2R \cdot \frac{1}{2}(2R - \mathrm{Crd}(180° - \alpha)) = R \cdot (2R - \mathrm{Crd}(180° - \alpha)), \tag{3.7}$$

proving the "second premise".

Algebraic Operations The text of MS J_2 now resumes at the same point where J_1 returns to the isosceles-trapezium cyclic quadrilateral $KBJD$ in figure 3.1. Since the isosceles trapezium's diagonals KJ and BD are equal, $KJ \cdot BD = KJ^2$. Recall that its base KD is the known Crd 6°, while each of its other three sides equals the unknown $y = \mathrm{Crd}\,2°$. Therefore the "first premise" or Ptolemy's theorem implies that

$$y^2 + \mathrm{Crd}\,6° \cdot y = KJ^2. \tag{3.8}$$

Note also that the diagonal KJ is the chord of twice 2°. So we can substitute $\widehat{KJ}$ for α, the unknown y for $\mathrm{Crd}\,\alpha/2$, and 60 for R in the "second premise":

$$y^2 = 60 \cdot \left(120 - \mathrm{Crd}\left(180° - \widehat{KJ}\right)\right) \quad \text{or} \quad \mathrm{Crd}\left(180° - \widehat{KJ}\right) = 120 - \frac{y^2}{60}. \tag{3.9}$$

Squaring both sides gives

$$\left(\mathrm{Crd}\left(180° - \widehat{KJ}\right)\right)^2 = 14400 - \frac{240\,y^2}{60} + \frac{y^4}{3600}. \tag{3.10}$$

On the other hand, by Thales' theorem and the Pythagorean theorem,

$$\left(\mathrm{Crd}\left(180° - \widehat{KJ}\right)\right)^2 = 14400 - KJ^2. \tag{3.11}$$

Therefore

$$KJ^2 = -\frac{y^4}{3600} + \frac{240\,y^2}{60}. \tag{3.12}$$

The text now combines the above two expressions for KJ^2 and simplifies:

$$-\frac{1}{3600}\cdot y^4+\frac{240}{60}\cdot y^2+0\,y=0\,y^4+1\,y^2+\mathrm{Crd}\,6^\circ\cdot y$$
$$3\,y^2=\frac{1}{3600}\cdot y^4+\mathrm{Crd}\,6^\circ\cdot y$$
$$10800\,y^2=1\,y^4+\mathrm{Crd}\,6^\circ\cdot 3600\,y$$
$$(3,0,0)\,y=y^3+(1,0,0)\,\mathrm{Crd}\,6^\circ$$
$$(3,0,0)\,y=y^3+6,16,49;7,59,8,56,29,40. \tag{3.13}$$

Iterative Algorithm Beginning with the immediately preceding equation in the form

$$y=\frac{6,16,49;7,59,8,56,29,40+y^3}{3,0,0}, \tag{3.14}$$

we can represent the prescribed iteration using notation and steps similar to those employed in the iteration for Y discussed in section 3.1. Namely, a *paṅkti* or current approximate value $y\approx q_0;q_1,q_2,\ldots,q_k$ is extracted digit by digit, with each digit q_i and corresponding "remainder" term r_i computed from the ones used in the preceding iteration. The first quotient digit q_0 and its remainder r_0 are defined as follows:

$$q_0=(6,16,49;7,59,8,56,29,40)\div(3,0,0)\qquad =2$$
$$r_0=6,16,49;7,59,8,56,29,40-(3,0,0)q_0\qquad =16,49;7,59,8,56,29,40.$$

The iteration process involves the following steps, with sexagesimal place-value adjusted in the operations to keep the digit sequence consistent:

- The y^3 term in the cubic equation for y is first roughly approximated by the cube of q_0 itself.
- The cube of q_0 (or strictly speaking, the cube difference $(q_0)^3-0^3$) is added to the remainder r_0, producing a sum that for convenience we denote p_1.
- The next digit q_1 is extracted by dividing p_1 by 3, 0, 0.
- The division remainder r_1 is added to the cube difference $(q_0;q_1)^3-(q_0)^3$, producing sum p_2.
- The next digit q_2 is extracted by dividing p_2 by 3, 0, 0, and so on.

More formally, we can define for consistency's sake a "leading zero" digit $q_{(-1)}=0$ as the starting-point for the algorithm. Then the kth sexagesimal digit q_k of the *paṅkti*, along with its corresponding remainder r_k, can be determined using the following operation cycle for $k\in\{1,2,3,\ldots\}$:

$$p_k=r_{k-1}+(0,q_0;q_1,q_2,\ldots,q_{k-1})^3-(0,q_0;q_1,q_2,\ldots,q_{k-2})^3 \tag{3.15}$$
$$q_k=((1,0)^k\cdot p_k)\div(3,0,0) \tag{3.16}$$
$$r_k=p_k-3\,q_k\cdot(1,0)^{2-k} \tag{3.17}$$

3.3.2 Calculations

The following tabular layouts summarize the steps of the nine computation cycles recorded in the manuscript, using the data values and notation supplied above, beginning with the "leading zero" digit $q_{(-1)}=0$. Note that starting in cycle 5, the author of these calculations truncates final digits of the sexagesimal cubes.

k	p_k	q_k, r_k, and *paṅkti* $q_0; q_1, q_2, \ldots, q_k$
1	$(q_{(-1)})^3 = 0^3 \quad = 0$ $(q_0)^3 = 2^3 \quad = 8$ 8 $-\ 0$ 8 $r_0:\ +16,49;7,59,8,56,29,40$ $p_1 =\ 16,57;7,59,8,56,29,40$	$q_1 = ((1,0)^1 \cdot p_1) \div (3,0,0)$ $= 16 \div 3 \quad = 5$ $r_1 = p_1 - 3\,q_1 \cdot (1,0)^1$ $= p_1 - 15,0$ $= 1,57;7,59,8,56,29,40$ *paṅkti*: 2;5
2	$(q_0;q_1)^3 = (2;5)^3 \quad = 9;2,32,5$ $9;\ 2,32,\ 5$ $-\ 8$ $1;\ 2,32,\ 5$ $r_1:\ +1,57;\ 7,59,\ 8,56,29,40$ $p_2 =\ 1,58;10,31,13,56,29,40$	$q_2 = ((1,0)^2 \cdot p_2) \div (3,0,0)$ $= 1,58 \div 3 \quad = 39$ $r_2 = p_2 - 3\,q_2 \cdot (1,0)^0$ $= p_2 - 1,57$ $= 1;10,31,13,56,29,40$ *paṅkti*: 2;5,39
3	$(q_0;q_1,q_2)^3 = (2;5,39)^3$ $= 9;11,2,32,27,43,39$ $9;11,\ 2,32,27,43,39$ $-9;\ 2,32,\ 5$ $0;\ 8,30,27,27,43,39$ $r_2:\ +1;10,31,13,56,29,40$ $p_3 =\ 1;19,\ 1,41,24,13,19$	$q_3 = ((1,0)^3 \cdot p_3) \div (3,0,0)$ $= 1,19 \div 3 \quad = 26$ $r_3 = p_3 - 3\,q_3 \cdot (1,0)^{-1}$ $= p_3 - 1;18$ $= 0;1,1,41,24,13,19$ *paṅkti*: 2;5,39,26
4	$(q_0;q_1,q_2,q_3)^3 = (2;5,39,26)^3$ $= 9;11,8,14,33,12,23,21,4,56$ $9;11,8,14,33,12,23,21,4,56$ $-9;11,2,32,27,43,39$ $0;\ 0,5,42,\ 5,28,44,21,4,56$ $r_3:\ +0;\ 1,1,41,24,13,19,\ 0,0,\ 0$ $p_4 =\ 0;\ 1,7,23,29,42,\ 3,21,4,56$	$q_4 = ((1,0)^4 \cdot p_4) \div (3,0,0)$ $= 1,7 \div 3 \quad = 22$ $r_4 = p_4 - 3\,q_4 \cdot (1,0)^{-2}$ $= p_4 - 0;1,6$ $= 0;0,1,23,29,42,3,21,4,56$ *paṅkti*: 2;5,39,26,22
5	$(q_0;q_1,q_2,q_3,q_4)^3 = (2;5,39,26,22)^3$ $\approx 9;11,8,19,22,41,6,52,15,1,56$ $9;11,8,19,22,41,\ 6,52,15,\ 1,56$ $-9;11,8,14,33,12,23,21,\ 4,56$ $0;\ 0,0,\ 4,49,28,43,31,10,\ 5,56$ $r_4:\ +0;\ 0,1,23,29,42,\ 3,21,\ 4,56,\ 0$ $p_5 =\ 0;\ 0,1,28,19,10,46,52,15,\ 1,56$	$q_5 = ((1,0)^5 \cdot p_5) \div (3,0,0)$ $= 1,28 \div 3 \quad = 29$ $r_5 = p_5 - 3\,q_5 \cdot (1,0)^{-3}$ $= p_5 - 0;0,1,27$ $= 0;0,0,1,19,10,46,52,15,1,56$ *paṅkti*: 2;5,39,26,22,29

k	p_k	q_k, r_k, and *paṅkti* $q_0; q_1, q_2, \ldots, q_k$
6	$(q_0; q_1, q_2, q_3, q_4, q_5)^3 = (2;5,39,26,22,29)^3$ $\approx 9;11,8,19,29,2,42,1,39,50,52$ $9;\ 11,8,19,29,\ 2,42,\ 1,39,50,52$ $-9;\ 11,8,19,22,41,\ 6,52,15,\ 1,56$ $0;\ 0,0,\ 0,\ 6,21,35,\ 9,24,48,56$ $r_5:\ +0;\ 0,0,\ 1,19,10,46,52,15,\ 1,56$ $p_6 =\ 0;\ 0,0,\ 1,25,32,22,\ 1,39,50,52$	$q_6 = ((1,0)^6 \cdot p_6) \div (3,0,0)$ $= 1,25 \div 3 \quad = 28$ $r_6 = p_6 - 3\,q_6 \cdot (1,0)^{-4}$ $= p_6 - 0;0,0,1,24$ $= 0;0,0,0,1,32,22,1,39,50,52$ *paṅkti*: $2;5,39,26,22,29,28$
7	$(q_0; q_1, q_2, q_3, q_4, q_5, q_6)^3 = (2;5,39,26,22,29,28)^3$ $\approx 9;11,8,19,29,8,50,27,19,59,43$ $9;\ 11,8,19,29,\ 8,50,27,19,59,43$ $-9;\ 11,8,19,29,\ 2,42,\ 1,39,50,52$ $0;\ 0,0,\ 0,\ 0,\ 6,\ 8,25,40,\ 8,51$ $r_6:\ +0;\ 0,0,\ 0,\ 1,32,22,\ 1,39,50,52$ $p_7 =\ 0;\ 0,0,\ 0,\ 1,38,30,27,19,59,43$	$q_7 = ((1,0)^7 \cdot p_7) \div (3,0,0)$ $= 1,38 \div 3 \quad = 32$ $r_7 = p_7 - 3\,q_7 \cdot (1,0)^{-5}$ $= p_7 - 0;0,0,0,1,36$ $= 0;0,0,0,0,2,30,27,19,59,43$ *paṅkti*: $2;5,39,26,22,29,28,32$
8	$(q_0; q_1, q_2, q_3, q_4, q_5, q_6, q_7)^3 = (2;5,39,26,22,29,28,32)^3$ $\approx 9;11,8,19,29,8,57,28,23,37,1$ $9;\ 11,8,19,29,8,57,28,23,37,\ 1$ $-9;\ 11,8,19,29,8,50,27,19,59,43$ $0;\ 0,0,\ 0,\ 0,0,\ 7,\ 1,\ 3,37,18$ $r_7:\ +0;\ 0,0,\ 0,\ 0,2,30,27,19,59,43$ $p_8 =\ 0;\ 0,0,\ 0,\ 0,2,37,28,23,37,\ 1$	$q_8 = ((1,0)^8 \cdot p_8) \div (3,0,0)$ $= 2,37 \div 3 \quad = 52$ $r_8 = p_8 - 3\,q_8 \cdot (1,0)^{-6}$ $= p_8 - 0;0,0,0,2,36$ $= 0;0,0,0,0,0,1,28,23,37,1$ *paṅkti*: $2;5,39,26,22,29,28,32,52$
9	$(q_0; q_1, q_2, q_3, q_4, q_5, q_6, q_7, q_8)^3$ $= (2;5,39,26,22,29,28,32,52)^3$ $\approx 9;11,8,19,29,8,57,39,47,50,19$ $9;\ 11,8,19,29,8,57,39,47,50,19$ $-9;\ 11,8,19,29,8,57,28,23,37,\ 1$ $0;\ 0,0,\ 0,\ 0,0,\ 0,11,24,13,18$ $r_8:\ +0;\ 0,0,\ 0,\ 0,0,\ 1,28,23,37,\ 1$ $p_9 =\ 0;\ 0,0,\ 0,\ 0,0,\ 1,39,47,50,19$	$q_9 = ((1,0)^9 \cdot p_9) \div (3,0,0)$ $= 1,39 \div 3 \quad = 33$ $r_9 = p_9 - 3\,q_9 \cdot (1,0)^{-7}$ $= p_9 - 0;0,0,0,1,39$ $= 0;0,0,0,0,0,0,0,47,50,19$ *paṅkti*: $2;5,39,26,22,29,28,32,52,33$

3.4 Muḥammad Ābida's Successive-Values Iteration for y, §2.4

3.4.1 Methods

This fixed-point iteration represents a trivial variant of the successive-values iterative equation for $y = \mathrm{Crd}\,2°$ seen in equation 1.5, namely:

$$y_{k+1} = y_0 + \frac{(y_k)^3}{3,0,0} \quad \text{with} \quad y_0 = \frac{(1,0,0)\,\mathrm{Crd}\,6°}{3,0,0} \approx \frac{6,16,49;7,59,8,56,30}{3,0,0} = 2;5,36,22,39,42,58,50 \qquad (3.18)$$

3.4.2 Calculations

The following table summarizes the five cycles of y_k iteration recorded in the manuscript following the statement of the y_0 value, with accompanying notes on the calculations.

k	y_k	Notes
1	$(y_0)^3 = (2;5,36,22,39,42,58,50)^3$ $\approx 9;10,28,3,8,52,5,39$ $\frac{(y_0)^3}{3,0,0} \approx 0;0,3,3,29,21,2,57$ $y_0:\quad 2;5,36,22,39,42,58,50$ $+0;0,3,3,29,21,2,57$ $y_1 = \quad 2;5,39,26,9,4,1,47$	Compare $\frac{(y_0)^3}{3,0,0} = \frac{9;10,28,3,8,52,5,38,21,47,43,\ldots}{3,0,0}$ $= 0;0,3,3,29,21,2,57,21,52,47,\ldots$ The calculator's upward rounding of the final cube digit does not affect the accuracy of the result at this precision.
2	$(y_1)^3 = (2;5,39,26,9,4,1,47)^3$ $\approx 9;11,8,16,32,30,48,9$ $\frac{(y_1)^3}{3,0,0} \approx 0;0,3,3,42,45,30,50$ $y_0:\quad 2;5,36,22,39,42,58,50$ $+0;0,3,3,42,45,30,50$ $y_2 = \quad 2;5,39,26,22,28,29,40$	Compare $\frac{(y_1)^3}{3,0,0} = \frac{9;11,8,16,32,30,47,49,12,13,55,33,\ldots}{3,0,0}$ $= 0;0,3,3,42,45,30,50,15,56,24,\ldots$ The calculator's last two cube digits 48, 9 should have been 47, 49, but since he evidently halted the division at the preceding cube digit 30, the discrepancy did not affect the result.
3	$(y_2)^3 = (2;5,39,26,22,28,29,40)^3$ $\approx 9;11,8,19,28,36,23,50$ $\frac{(y_2)^3}{3,0,0} \approx 0;0,3,3,42,46,29,39$ $y_0:\quad 2;5,36,22,39,42,58,50$ $+0;0,3,3,42,46,29,39$ $y_3 = \quad 2;5,39,26,22,29,28,29$	Compare $\frac{(y_2)^3}{3,0,0} = \frac{9;11,8,19,28,56,2,53,51,21,20,\ldots}{3,0,0}$ $= 0;0,3,3,42,46,29,38,40,57,57,\ldots$ The cube digit 36, shown in red, was apparently written in error for 56, but the accurate result of the cube's subsequent division by 3, 0, 0 implies that the calculator's dividend was correctly written with 56. Perhaps the scribal error resulted from misinterpreting an *abjad* number in a Persian or Arabic source text: i.e., *abjad* نو 56 was misread as لو 36.
4	$(y_3)^3 = (2;5,39,26,22,29,28,29)^3$ $\approx 9;11,8,19,29,8,56,51$ $\frac{(y_3)^3}{3,0,0} \approx 0;0,3,3,42,46,29,42$ $y_0:\quad 2;5,36,22,39,42,58,50$ $+0;0,3,3,42,46,29,42$ $y_4 = \quad 2;5,39,26,22,29,28,32$	Compare $\frac{(y_3)^3}{3,0,0} = \frac{9;11,8,19,29,8,56,48,55,9,9,30,\ldots}{3,0,0}$ $= 0;0,3,3,42,46,29,42,58,56,\ldots$ Again, the slight inaccuracy in the least significant cube digit does not affect the result of the division by 3, 0, 0. But the quotient from that division was evidently truncated to 42 rather than rounded up to 43 in the last place.
5	$(y_4)^3 = (2;5,39,26,22,29,28,32)^3$ $\approx 9;11,8,19,29,8,57,28$ $\frac{(y_4)^3}{3,0,0} \approx 0;0,3,3,42,46,29,42$ $y_0:\quad 2;5,36,22,39,42,58,50$ $+0;0,3,3,42,46,29,42$ $y_5 = \quad 2;5,39,26,22,29,28,32$	Compare $\frac{(y_4)^3}{3,0,0} = \frac{9;11,8,19,29,8,57,28,23,37,1,53,\ldots}{3,0,0}$ $= 0;0,3,3,42,46,29,42,59,9,\ldots$ Using these more precise figures, the final value would converge to 2; 5, 39, 26, 22, 29, 28, 33 in the seventh sexagesimal place.

3.5 Ulugh Beg's Successive-Values Iteration for y, §2.5

3.5.1 Methods

This material, attributed in both manuscript recensions to Ulugh Beg, begins with another theoretical derivation of the cubic equation for $y = \mathrm{Crd}\,2°$, followed by a successive-values technique to determine its value. Again following the notation of MS J_1, we consider the new, mirror-image diagram of cyclic quadrilateral $KBJD$ shown in figure 3.3.

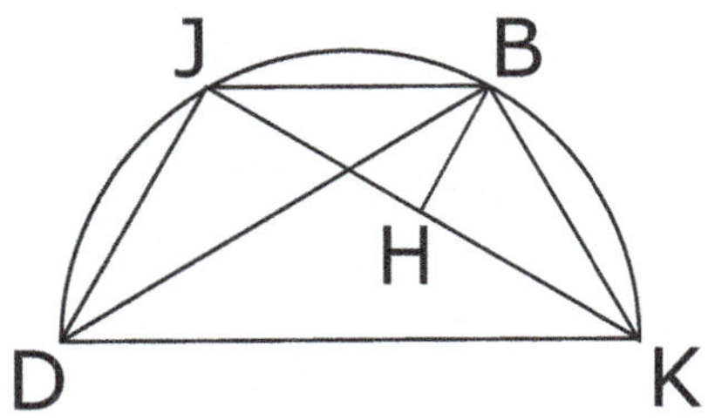

Figure 3.3: Diagram reproduced from figure J_1.3

A perpendicular is dropped from point B onto H on diagonal KJ. Since triangles KBH and JBH are right,

$$KB^2 = KH^2 + BH^2 \quad \text{and} \quad BJ^2 = JH^2 + BH^2;$$

but BH is common to both triangles, and $KB = BJ$. Therefore the triangles are congruent, and $KH = HJ$.

The author claims, correctly but without proof, that

$$KB^2 = 2R \cdot BH, \quad \text{and so} \quad \frac{KB^2}{2R} = BH. \tag{3.19}$$

MS J_2 notes that this equality results from the 8th and 17th "figures", i.e., propositions, in the sixth chapter of *Rekhāgaṇita* or *Elements*. The reasoning can be reconstructed as follows, with reference to figure 3.4: Extend BH through the center of the circle to point W opposite, producing $BW = 2R$, and join WK. Then by Thales' theorem, triangle BKW is right; and since KH is perpendicular to BH, by *Elements* VI.8 triangle BKW is similar to triangle BHK, so $BH/KB = KB/BW$. Then, by *Elements* VI.17, $KB^2 = BW \cdot BH = 2R \cdot BH$.

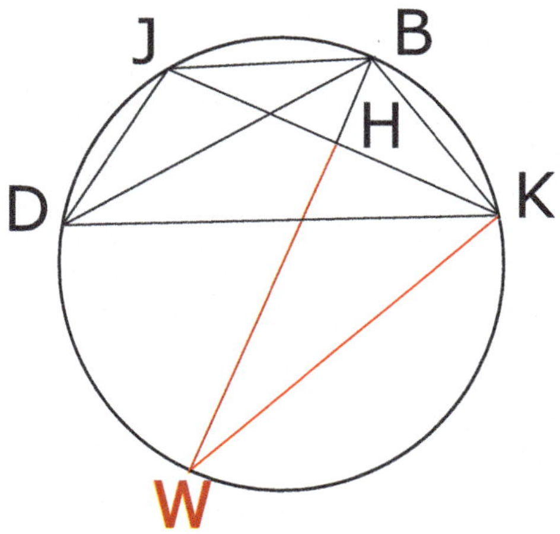

Figure 3.4: Diagram modified from figure J_1.3

Recalling that KB is the unknown y and that by the Pythagorean theorem $KB^2 - BH^2 = KH^2$, it follows from the above demonstration that

$$KH^2 = KB^2 - \left(\frac{KB^2}{2R}\right)^2 = y^2 - \left(\frac{y^2}{2,0}\right)^2 = y^2 - (0;0,0,15)\,y^4. \tag{3.20}$$

Since KH is half of chord KJ, KH^2 is one-fourth of KJ^2, so

$$KJ^2 = 4\,KH^2 = 4y^2 - (0;0,1)\,y^4. \tag{3.21}$$

And by Ptolemy's theorem (*Almagest* 1.2) as previously explained,

$$KJ^2 = KB^2 + KB \cdot \text{Crd}\,6^\circ = y^2 + y \cdot \text{Crd}\,6^\circ, \tag{3.22}$$

so

$$\begin{aligned}
4y^2 - (0;0,1)\,y^4 &= y^2 + y \cdot \text{Crd}\,6^\circ \\
3y^2 &= (0;0,1)\,y^4 + y \cdot \text{Crd}\,6^\circ \\
y^2 &= (0;0,0,20)\,y^4 + y \cdot \frac{\text{Crd}\,6^\circ}{3} \\
y &= (0;0,0,20)\,y^3 + \frac{\text{Crd}\,6^\circ}{3} \\
&= (0;0,0,20)\,y^3 + 2;5,36,22,39,42,58,50,
\end{aligned} \tag{3.23}$$

neglecting the inaccuracy of the constant term.

This cubic equation in y, like the previously derived ones, could not be solved by ordinary algebra, but we can generate the desired value of y by iteration. The computation is, once again, equivalent to iterating the y version of equation 1.5:

$$y_{k+1} = y_0 + (0;0,0,20)\,(y_k)^3 \quad \text{with} \quad y_0 = \frac{\text{Crd}\,6^\circ}{3} \approx 2;5,36,22,39,42,58,50. \tag{3.24}$$

3.5.2 Calculations

Four repetitions of this iteration cycle yield the same fixed value as in the computations of section 3.4.2: $y_4 = 2;5,39,26,22,29,28,32$.

The steps of the procedure are likewise identical to those in the previous section's calculations of y_1 through y_4, except that here the operation of division by 3, 0, 0 is described as multiplying by 20 thirds or $20'''$. The only difference in the numerical data is the more accurate statement of the $(y_2)^3$ value, containing the correct sexagesimal digit 56 rather than 36 in the fifths place.

3.6 Ulugh Beg's Successive-Values Iteration for x, §2.6

3.6.1 Methods

This "second method" attributed to Ulugh Beg, for approximating $x = R \sin 1^\circ$, begins with a more elaborate version of the usual isosceles-trapezium diagram, which we discuss here in its MS J_1 version (on left in figure 3.5, which also provides for comparison the variants in MS J_2 and in the Persian commentary of al-Bīrjandī). The circumcircle of the trapezium KB_1JD has radius length R and center T, with points B_1 and J dividing the six-degree arc KD into equal thirds as before. With chords connecting each pair among those points, circumcircle radii drawn from T will perpendicularly bisect chord KB_1 at H, chord KJ at Jh, and chord KD at B_2. All the resulting right triangles THK, $TJhK$, TB_2K can be inscribed in a semicircle whose diameter is their common hypotenuse $KT = R$, which we bisect at A.

Joining chords HJh, JhB_2, and B_2T in this smaller semicircle TB_2JhHK (the text also repeats the direction to draw perpendicular bisector HT), we produce triangle $KJhB_2$ similar to KJD, with side lengths one-half those of the larger triangle. (We can see that in the same way, triangle $KHJh$ is a half-scale version of triangle KB_1J, and KHB_2 of KB_1D: in fact, the entire small cyclic quadrilateral $KHJhB_2$ is just a half-scale version of the original KB_1JD.)

The author also notes that radius ALH is the perpendicular bisector of KJh. But the central fact is that chord KH is half of KB_1 or Crd 2°, so $KH = R \sin 1^\circ = x$, while $KB_2 = \frac{1}{2}\,\text{Crd}\,6^\circ \approx 3;8,24,33,59,34,28,15$.

The relations among the segments KH, KJh and HL can be turned into symbolic equations in the unknown $KH = x$. Although the author does not explicitly make this argument, we can invoke the same Euclidean propositions we used when

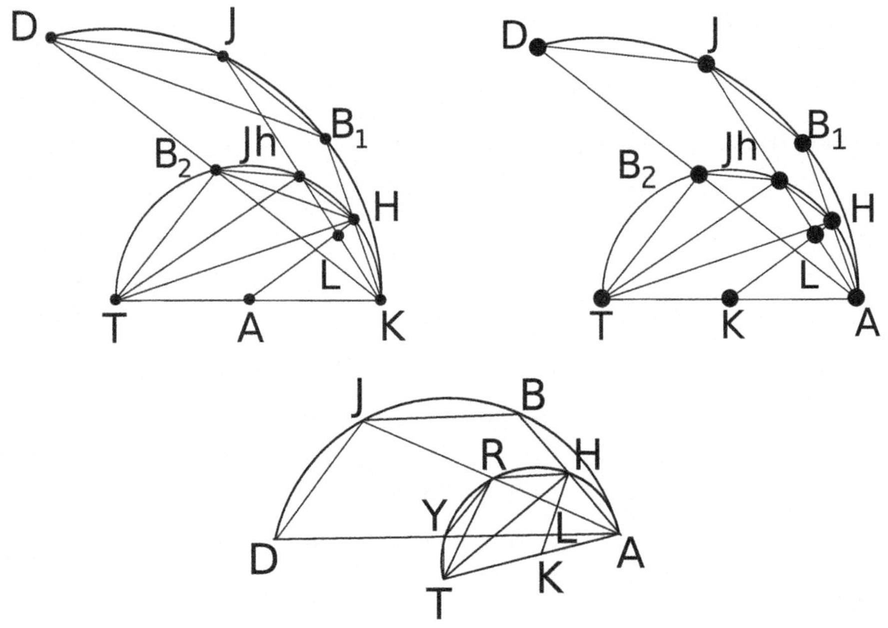

Figure 3.5: Versions of the diagram illustrating Ulugh Beg's second method. Left: Reproduced from figure J_1.4. Right: reproduced from figure J_2.3. Bottom: reproduced from al-Bīrjandī's version in figure 4.9. The *nāgarī* letter labels on points in the Sanskrit manuscript diagrams deviate somewhat from their counterparts in the Persian original. The Persian letter *ya* or Y has been replaced by a second instance of the *nāgarī* letter *ba*; we have added subscripts 1 and 2 in order to distinguish between the two points B in the translated versions. And Persian *ra* or R is represented by *nāgarī jha*, the aspirated voiced palatal consonant, which we denote by Jh.

discussing figure 3.4 in section 3.5.1 to show that

$$x^2 = KH^2 = 2\,(AH) \cdot HL = R \cdot HL, \quad \text{whence} \quad HL^2 = \frac{(x^2)^2}{R} = (0;0,1)x^4. \tag{3.25}$$

The Pythagorean relation and the fact that KL is half of KJh give

$$\begin{aligned} \frac{1}{4}KJh^2 &= KL^2 = KH^2 - HL^2 = x^2 - (0;0,1)x^4 \\ KJh^2 &= 4x^2 - (0;0,4)x^4. \end{aligned} \tag{3.26}$$

"Furthermore" (by Ptolemy's theorem),

$$KJh^2 = KH^2 + KH \cdot R\sin 3° = x^2 + x \cdot R\sin 3°; \tag{3.27}$$

so setting these two expressions for KJh^2 equal, we derive

$$\begin{aligned} 4x^2 - (0;0,4)\,x^4 &= x^2 + x \cdot R\sin 3° \\ \Longrightarrow \qquad x &= (0;0,1,20)\,x^3 + 1;2,48,11,19,51,29,25, \end{aligned} \tag{3.28}$$

again neglecting the inaccuracy of the constant term.

3.6.2 Calculations

The fixed-point iterative algorithm based on the foregoing equation is equivalent to the x-version of equation 1.5:

$$x_{k+1} = x_0 + (0;0,1,20)\,(x_k)^3 \qquad \text{with} \qquad x_0 = \frac{R\sin 3°}{3} = 1;2,48,11,19,51,29,25. \tag{3.29}$$

The following table summarizes the four cycles of x_k iteration carried out in the manuscript calculations, with accompanying notes.

k	x_k	Notes
1	$(x_0)^3 = (1;2,48,11,19,51,29,25)^3$ $\approx 1;8,48,30,23,36,30,42,18$ $(x_0)^3(0;0,1,20) \approx 0;0,1,31,44,40,31,28,41$ $x_0:$ $1;2,48,11,19,51,29,25$ $+0;0,1,31,44,40,31,28,41$ $x_1 =$ $1;2,49,43,4,32,0,53,41$	Compare $(x_0)^3 = 1;8,48,30,23,36,30,42,17,43,\ldots$ $(x_0)^3(0;0,1,20)$ $= 0;0,1,31,44,40,31,28,40,56,\ldots$
2	$(x_1)^3 = (1;2,49,43,4,32,0,53,41)^3$ $\approx 1;8,53,32,4,3,50,59,14,57$ $(x_1)^3(0;0,1,20) \approx 0;0,1,31,51,22,45,25,8$ $x_0:$ $1;2,48,11,19,51,29,25$ $+0;0,1,31,51,22,45,25,8$ $x_2 =$ $1;2,49,43,11,14,14,50,8$	Compare $(x_1)^3 = 1;8,53,32,4,3,50,59,15,12,49,\ldots$ $(x_1)^3(0;0,1,20)$ $= 0;0,1,31,51,22,45,25,7,59,0,\ldots$ The erroneous two final digits of the cube of x_1 as stated in the manuscript do not affect the accuracy of the result of the multiplication.
3	$(x_2)^3 = (1;2,49,43,11,14,14,50,8)^3$ $\approx 1;8,53,36,26,7,0,22,10$ $(x_2)^3(0;0,1,20) \approx 0;0,1,31,51,23,14,49,20$ $x_0:$ $1;2,48,11,19,51,29,25$ $+0;0,1,31,51,23,14,49,20$ $x_3 =$ $1;2,49,43,11,14,44,14,20$	Compare $(x_2)^3 = 1;8,53,32,26,7,0,22,10,14,\ldots$ $(x_2)^3(0;0,1,20)$ $= 0;0,1,31,51,23,14,49,20,29,\ldots$ Again, the value of $(x_2)^3$ used in the actual calculations was evidently more accurate than the number recorded in the manuscript recensions, since the error of substituting 36 for 32 in one digit did not propagate through subsequent results.
4	$(x_3)^3 = (1;2,49,43,11,14,44,14,20)^3$ $\approx 1;8,53,32,26,8,37,5,34$ $(x_3)^3(0;0,1,20) \approx 0;0,1,31,51,23,14,51,29$ $x_0:$ $1;2,48,11,19,51,29,25$ $+0;0,1,31,51,23,14,51,29$ $x_4 =$ $1;2,49,43,11,14,44,16,26$	Compare $(x_3)^3 = 1;8,53,32,26,8,37,5,33,59,\ldots$ $(x_3)^3(0;0,1,20)$ $= 0;0,1,31,51,23,14,51,29,27,\ldots$ The 26 in lieu of 29 in the last digit of the final result is apparently another scribal error.

3.7 Ābida's Alternate Proof of Ulugh Beg's Method for Generating a Cubic Equation for x, §2.7

3.7.1 Methods

The geometric derivation of the cubic equation for $x = R \sin 1°$ in section 2.6 was apparently regarded by the author of our text as unnecessarily cumbersome compared to the demonstration that he now describes, attributing it to Muḥammad Ābida. Instead of the elaborate configuration of two similar cyclic quadrilaterals depicted in figure 3.5, only the familiar quadrilateral $KBJD$ is employed. As shown at left in figure 3.6, it now includes an additional point H whose location is not very clearly demarcated in the manuscript sketches, but according to the text is evidently the midpoint of chord BK. Since $KB = BJ = JD = \mathrm{Crd}\, 2° = y$ as before, the half-chord KH is therefore equal to $y/2 = x = R \sin 1°$.

The length of the base $KD = \mathrm{Crd}\, 6°$ is stated as 6; 16, 49, 7, 59, 8, 16. Note that according to the standard value of this constant throughout the text, the sexagesimal digit in its sixth fractional place ought to be 56 instead of 16. As in a different exposition of Ābida's work in section 2.4 where a similar numerical error appeared, this confusion between 16 and 56 may have been the result of misreading *abjad* نو , or 56, as *abjad* يو , or 16.

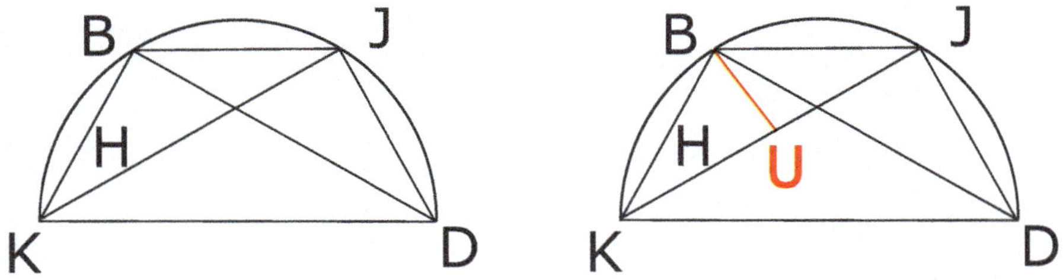

Figure 3.6: Left: Diagram reproduced from figure J_1.5. Right: Modified version of same diagram.

By Ptolemy's theorem,

$$\begin{aligned} KB \cdot JD + KD \cdot BJ &= 2x \cdot 2x + (6; 16, 49, 7, 59, 8, 16) \cdot 2x \\ &= 4x^2 + (12; 33, 38, 15, 58, 16, 33, 0)\, x \\ &= KJ \cdot BD \\ &= KJ^2. \end{aligned} \tag{3.30}$$

The erroneous final digit 16 of Crd 6° (presumably in combination with some truncated fraction) evidently was doubled to yield the final digits 33, 0 appearing in 2 Crd 6° = 12; 33, 38, 15, 58, 16, 33, 0.

The "other method" for determining the square of KJ introduces, without explanatory derivation, a quantity not shown in the original diagram: the R versine of 2°, equal to $R - R\cos 2°$. That R vers 2° is represented in our modified version of the diagram at right in figure 3.6 by the length of the perpendicular BU dropped from B onto KJ = Crd 4°.

Again invoking the Euclidean rationale illustrated by figure 3.4 in section 3.5.1, we can state that

$$R \operatorname{vers} 2° = \frac{KB^2}{2R} = \frac{(2x)^2}{2,0} = (0; 2)\, x^2. \tag{3.31}$$

Then because the perpendicular $BU = R$ vers 2° bisects chord KJ, the Pythagorean theorem implies that

$$\left(\frac{KJ}{2}\right)^2 = KB^2 - (R \operatorname{vers} 2°)^2 = 4x^2 - (0; 0, 4)x^4, \tag{3.32}$$

so

$$KJ^2 = 16x^2 - (0; 0, 16)x^4. \tag{3.33}$$

Equating the above two expressions for KJ^2 yields the same cubic equation in x employed in section 3.6.1:

$$\begin{aligned} 4x^2 + (12; 33, 38, 15, 58, 16, 33, 0)x &= 16x^2 - (0; 0, 16)x^4 \\ 12x^2 &= (0; 0, 16)x^4 + (12; 33, 38, 15, 58, 16, 33, 0)x \\ \Longrightarrow \qquad x &= (0; 0, 1, 20)\, x^3 + 1; 2, 48, 11, 19, 51, 29, 25. \end{aligned} \tag{3.34}$$

However, the actual number that was divided by 12 to produce the constant value 1; 2, 48, 11, 19, 51, 29, 25 in the final cubic equation must instead have been 12; 33, 38, 15, 58, 17, 52 = 2 · 6; 16, 49, 7, 59, 8, 56—i.e., twice the *correct* value of Crd 6°.

3.7.2 Calculations

Again, the immediately preceding cubic equation for x corresponds to the x-version of iterative equation 1.5:

$$x_{k+1} = 1; 2, 48, 11, 19, 51, 29, 25 + (0; 0, 1, 20)\,(x_k)^3, \tag{3.35}$$

although no computations are presented in the text.

3.8 Dayānat Khān's Successive-Digits Iteration for x, §2.8

3.8.1 Methods

The method explained in the last part of J_2, where it is ascribed to Dayānat Khān, begins by reducing the same quartic equation that appeared most recently near the end of the preceding section 3.7.1. The author re-uses the same erroneous 16, 33 instead of 17, 52 in the final digits of the equation's constant term, writing some of the coefficients in arcseconds notation rather than as sexagesimal place-value fractions:

$$\begin{aligned} 12x^2 &= 16''x^4 + (12;33,38,15,58,16,33)x \\ 3x &= 4''x^3 + 3;8,24,33,59,34,28,15 \end{aligned} \tag{3.36}$$

In a somewhat opaque way that might be an attempt to express a collaborator's explanation of some feature of Islamic algebra, the author describes in his "demonstration" an implied sum of terms on the left side: "Since in the second side there is a pair of quantities, hence in the first side too a pair of quantities is presumed, because of its equality". He appears to mean that the equation could be trivially rewritten as follows:

$$4''x^3 + \left(3x - 4''x^3\right) = 4''x^3 + 3;8,24,33,59,34,28,15 \tag{3.37}$$

From this left-side sum we can factor out x, yielding addends that he describes as "the product of the square of the produced number and four seconds" and the more obscurely expressed "as much as with which, being multiplied by the first number, is equal to the amount of the constant" (!):

$$x\left(4''x^2 + \left(3 - 4''x^2\right)\right) = 4''x^3 + 3;8,24,33,59,34,28,15 \tag{3.38}$$

The equivalent form that gives rise to the actual calculations is

$$x\left(3 - (0;0,4)x^2\right) = 3;8,24,33,59,34,28,15,\ldots \tag{3.39}$$

The method is a fairly complicated recursion based upon treating the desired sine value x as the sum of a known sexagesimal digit d plus an unknown remainder s, which itself is the sum of a digit and a remainder, and so on:

$$x = d_0 + s_0 = d_0 + (d_1 + s_1) = d_0 + d_1 + (d_2 + s_2) = \ldots = d_0 + d_1 + d_2 + \ldots + (d_k + s_k). \tag{3.40}$$

The two parts of the digit-and-remainder sum in each recursion cycle are algebraically distributed over the cubic equation to facilitate computation. We illustrate the basic structure of the method by assigning

$$x = 1 + s, \tag{3.41}$$

where the digit 1 is the result of integer division of $R \sin 3°$ by 3, and s is its sexagesimal-fraction remainder. Then the cubic equation can be rewritten as

$$3;8,24,33,59,34,28,15,\ldots = (1+s)\left(3 - 4''(1+s)^2\right), \tag{3.42}$$

which is manipulated in the following steps:

Step 1: Expand the square term in the second factor and distribute over its terms the 4″ multiplicand.

$$\begin{aligned} 3;8,24,33,59,34,28,15,\ldots &= (1+s)\left(3 - (0;0,4)(1+2s+s^2)\right) \\ &= (1+s)\left(3 - 0;0,4 - (0;0,4)2s - (0;0,4)s^2\right) \end{aligned}$$

Step 2: Break up the terms in the first factor $(1+s)$ for distribution over that expanded expression.

$$\begin{aligned} 3;8,24,33,59,34,28,15,\ldots = (1)&\left(3 - 0;0,4 - (0;0,4)2s - (0;0,4)s^2\right) \\ + s&\left(3 - (0;0,4) - (0;0,4)2s - (0;0,4)s^2\right) \end{aligned}$$

Step 3: Combine the constants in a single term.

$$3;8,24,33,59,34,28,15,\ldots = 2;59,56 + (1)\left(-(0;0,4)2s - (0;0,4)s^2\right) + s\left(2;59,56 - (0;0,4)2s - (0;0,4)s^2\right)$$

Step 4: Move the collected constant terms over to the constant side and multiply out the rest.

$$0;8,28,33,59,34,28,15,\ldots = -(0;0,4)2s - (0;0,4)s^2 + 2;59,56s - (0;0,4)(2)s^2 - (0;0,4)s^3$$

Step 5: Collect the linear terms on the right side. (The prescribed method leaves the two quadratic terms uncollected.)

$$0;8,28,33,59,34,28,15,\ldots = (2;59,48)s - (0;0,4)s^2 - (0;0,4)(2)s^2 - (0;0,4)s^3$$

The next task is to "generate" a new sexagesimal digit representing the approximate size of the unknown s by inspecting this equation. A more general simplified expression for the equation in the kth recursion cycle can be represented as follows:

$$\begin{aligned} c_k &= a_k s_k - b_k(s_k)^2 - (0;0,4)(s_k)^3 \\ &= s_k\left(a_k - b_k s_k - (0;0,4)(s_k)^2\right) \\ &= (d_{k+1} + s_{k+1})\left(a_k - b_k(d_{k+1} + s_{k+1}) - (0;0,4)(d_{k+1} + s_{k+1})^2\right) \end{aligned}$$

where a_k and b_k are the coefficients for the linear and (combined) quadratic terms respectively, c_k is the current collection of the constant terms, and d_{k+1} is the next sexagesimal digit in the result, selected so that the product $a_k \cdot d_{k+1}$ does not exceed the current remainder c_k.

We can think of the algebraic manipulations we just completed as an application of this equation to "pre-initial" values (using subscript (-1) as in section 3.3) consisting of the following:

- a supposed leading digit $d_{(-1)} = 0$,
- a "remainder" $s_{(-1)} = x$, which in turn is equal to the sum of $d_0 = 1$ and a remainder which we now designate s_0 rather than s,
- constant term $c_{(-1)} = 3;8,24,33,59,34,28,15,\ldots$,
- and coefficients $a_{(-1)} = 3$ and $b_{(-1)} = 0$.

We thus derive

$$\begin{aligned} c_{(-1)} &= a_{(-1)}s_{(-1)} - b_{(-1)}(s_{(-1)})^2 - (0;0,4)(s_{(-1)})^3 \\ 3;8,24,33,59,34,28,15,\ldots &= 3x - 0x^2 - (0;0,4)x^3 \\ &= x(3 - (0;0,4)x^2) \\ &= (1+s_0)(3 - (0;0,4)(1+s_0)^2) \\ &\vdots \\ 0;8,28,33,59,34,28,15,\ldots &= (2;59,48)s_0 - (0;0,12)(s_0)^2 - (0;0,4)(s_0)^3. \end{aligned}$$

These numbers give us the constant and coefficients for recursion cycle $k = 0$, to wit:

$$c_0 = 0;8,28,33,59,34,28,15,\ldots, \quad a_0 = 2;59,48, \quad b_0 = 0;0,12.$$

We can now move into recursion cycle $k = 1$ by breaking up s_0 into the sum of a sexagesimal digit d_1 and a new remainder s_1. By inspection, we determine that d_1 must be bigger than 0; 2 but less than 0; 3, so we rewrite our expressions as follows:

$$\begin{aligned} s_0 &= 0;2 + s_1 \\ 0;8,28,33,59,34,28,15,\ldots &= s_0\left(2;59,48 - (0;0,12)s_0 - (0;0,4)(s_0)^2\right) \\ &= (0;2+s_1)\left(2;59,48 - (0;0,12)(0;2+s_1) - (0;0,4)(0;2+s_1)^2\right) \end{aligned}$$

Stepping through the manipulation of this equation as in the previous cycle:

Step 1: Expand the expression in the second factor.

$$\begin{aligned}
c_0 &= (d_1 + s_1)\left(a_0 - b_0(d_1 + s_1) - (0;0,4)(d_1 + s_1)^2\right) \\
&= (d_1 + s_1)\left(a_0 - b_0 d_1 - b_0 s_1 - (0;0,4)(d_1)^2 - (0;0,4)2d_1 s_1 - (0;0,4)(s_1)^2\right) \\
0;8,28,33,59,34,28,15,\ldots &= (0;2 + s_1)\left(2;59,48 - (0;0,12)(0;2 + s_1) - (0;0,4)(0;2 + s_1)^2\right) \\
&= (0;2 + s_1)(2;59,48 - 0;0,0,24 - (0;0,12)s_1 \\
&\quad - (0;0,4)(0;0,4) - (0;0,4)(0;4)s_1 - (0;0,4)(s_1)^2) \\
&= (0;2 + s_1)(2;59,48 - 0;0,0,24 - (0;0,12)s_1 \\
&\quad - (0;0,0,0,16) - (0;0,0,16)s_1 - (0;0,4)(s_1)^2)
\end{aligned}$$

Step 2: Break up the terms in the first factor for distribution over that expanded expression.

$$\begin{aligned}
c_0 &= d_1\left(a_0 - b_0 d_1 - b_0 s_1 - (0;0,4)(d_1)^2 - (0;0,4)2d_1 s_1 - (0;0,4)(s_1)^2\right) \\
&\quad + s_1\left(a_0 - b_0 d_1 - b_0 s_1 - (0;0,4)(d_1)^2 - (0;0,4)2d_1 s_1 - (0;0,4)(s_1)^2\right) \\
0;8,28,33,59,34,28,15,\ldots &= (0;2)\,(2;59,48 - 0;0,0,24 - (0;0,12)s_1 \\
&\quad - (0;0,0,0,16) - (0;0,0,16)s_1 - (0;0,4)(s_1)^2\Big) \\
&\quad + s_1\,(2;59,48 - 0;0,0,24 - (0;0,12)s_1 \\
&\quad -(0;0,0,0,16) - (0;0,0,16)s_1 - (0;0,4)(s_1)^2\Big)
\end{aligned}$$

Step 3: Collect the constant terms separately, and the other like terms within each bracketed expression.

$$\begin{aligned}
c_0 &= d_1(a_0 - b_0 d_1 - (0;0,4)(d_1)^2) + d_1\left(-b_0 s_1 - (0;0,4)2d_1 s_1 - (0;0,4)(s_1)^2\right) \\
&\quad + s_1\left(a_0 - b_0 d_1 - (0;0,4)(d_1)^2 - b_0 s_1 - (0;0,4)2d_1 s_1 - (0;0,4)(s_1)^2\right) \\
0;8,28,33,59,34,28,15,\ldots &= (0;2)(2;59,48 - 0;0,0,24 - 0;0,0,0,16) \\
&\quad + (0;2)\left(-(0;0,12)s_1 - (0;0,0,16)s_1 - (0;0,4)(s_1)^2\right) \\
&\quad + s_1\,(2;59,48 - 0;0,0,24 - 0;0,0,0,16 \\
&\quad -(0;0,12)s_1 - (0;0,0,16)s_1 - (0;0,4)(s_1)^2\Big) \\
&= 0;5,59,35,11,28 + (0;2)\left(-(0;0,12,16)s_1 - (0;0,4)(s_1)^2\right) \\
&\quad + s_1\left(2;59,47,35,44 - (0;0,12,16)s_1 - (0;0,4)(s_1)^2\right)
\end{aligned}$$

Step 4: Move the collected constant terms over to the constant side and multiply out the rest.

$$\begin{aligned}
c_1 = c_0 - d_1(a_0 - b_0 d_{_} - (0;0,4)(d_1)^2) &= -b_0 d_1 s_1 - (0;0,4)(2)(d_1)^2 s_1 - (0;0,4)(d_1)(s_1)^2 \\
&\quad + a_0 s_1 - b_0 d_1 s_1 - (0;0,4)(d_1)^2 s_1 - b_0(s_1)^2 - (0;0,4)(2d_1)(s_1)^2 - (0;0,4)(s_1)^3 \\
0;2,28,58,48,6,28,15,\ldots &= -(0;0,0,24,32)s_1 - (0;0,0,8)(s_1)^2 \\
&\quad + (2;59,47,35,44)s_1 - (0;0,12,16)(s_1)^2 - (0;0,4)(s_1)^3
\end{aligned}$$

Step 5: Collect the linear terms on the right side.

$$\begin{aligned}
c_1 &= (a_0 - d_1(b_0 + (0;0,4)d_1) - d_1(b_0 + (0;0,4)(2d_1)))s_1 \\
&\quad - (b_0 + (0;0,4)d_1 + (0;0,4)(2d_1))(s_1)^2 - (0;0,4)(s_1)^3 \\
0;2,28,58,48,6,28,15,\ldots &= (2;59,47,11,12)s_1 - (0;0,12,24)(s_1)^2 - (0;0,4)(s_1)^3
\end{aligned}$$

The rather cumbersome algebraic expressions in Step 5 allow us to represent the new numerical values in the $(k+1)$th cycle, in terms of their previous values in the kth cycle, in a form that reflects their manipulation in the calculation:

New d for next sexagesimal digit of the result: The value d_{k+1} is chosen such that the product $a_k \cdot d_{k+1}$ does not exceed c_{k+1}.
New c for constant remainder: $c_{k+1} = c_k - d_{k+1}(a_k - b_k d_{k+1} - (0;0,4)(d_{k+1})^2)$
New a for linear term coefficient: $a_{k+1} = a_k - d_{k+1}(b_k + (0;0,4)d_{k+1}) - d_{k+1}(b_k + (0;0,4)2d_{k+1})$
New b for quadratic term coefficient: $b_{k+1} = b_k + (0;0,4)d_{k+1} + (0;0,4)2d_{k+1}$

It is the generation of these coefficients that is set out in the text and recorded in the table (see table 3.7) in the three distinct *paṅktis*. Along with them appears a related quantity here called g_k, such that

$$g_k = \frac{b_k}{(0;0,4)},$$

which is recorded in the third *paṅkti* instead of b, presumably because it is easier to handle. Note also that the expression for b_{k+1} above may be simplified easily by just adding its three terms. However, its current form allows a calculator to reuse the value of the term $2(0;0,4)d_{k+1}$ that had been found when calculating a_{k+1}.

3.8.2 Calculations

Here we recompute the successive coefficients in each recursion cycle (and the intermediate steps) and compare them to the quantitites recorded in the table (3.7). To do this, we have color-coded the successive values of these quantities, both in the calculations that follow and in the grid in figure 3.7, for each step of the iteration, with the first step in brown. Note that the base-60 place-value layout of the grid columns is not entirely consistent, but tends to shift one place to the right at each iteration.

The "pre-initial" cycle, as we described in the previous section, may be considered to assign $a_{(-1)} = 3$, $b_{(-1)} = 0$, $g_{(-1)} = 0$, $c_{(-1)} = 3;8,24,33,59,34,28,15$, and $d_0 = 1$. The initial cycle with $k = 0$ then produces the following:

$$\begin{aligned}
a_0 &= a_{(-1)} - d_0(b_{(-1)} + (0;0,4)d_{(-1)}) - d_0(b_{(-1)} + 2(0;0,4)d_0) \\
&= 3 - 1(0 + (0;0,4)\cdot 1) - 1(0 + 2(0;0,4)\cdot 1) \\
&= 3 - 0;0,4 - 0;0,8 \\
&= 2;59,56 - 0;0,8 \\
&= 2;59,48 \\
b_0 &= b_{(-1)} + (0;0,4)d_0 + 2(0;0,4)d_0 \\
&= 0 + (0;0,4)\cdot 1 + 2(0;0,4)\cdot 1) \\
&= 0;0,12 \\
g_0 &= g_{(-1)} + d_0 + 2d_0 \\
&= 0 + 1 + 2 \\
&= 3 \\
c_0 &= c_{(-1)} - (a_{(-1)}d_0 - b_{(-1)}(d_0)^2 - (0;0,4){d_0}^3) \\
&= 3;8,24,33,59,34,28,15 - (3\cdot 1 - 0\cdot 1^2 - (0;0,4)\cdot 1^3) \\
&= 3;8,24,33,59,34,28,15 - 2;59,56 = 0;8,28,33,59,34,28,15.
\end{aligned}$$

The value d_1 is then chosen to be the largest number of minutes so that $d_1 \cdot a_0$ does not exceed c_0, so $d_1 = 0;2$. The numbers in the second step are shown in red:

$$
\begin{aligned}
a_1 &= a_0 - d_1(b_0 + (0;0,4)d_1) - d_1(b_0 + 2(0;0,4)d_1)\\
&= 2;59,48 - (0;2)(0;0,12 + (0;0,4)(0;2)) - (0;2)(0;0,12 + 2(0;0,4)(0;2))\\
&= 2;59,48 - 0;0,0,24,16 - 0;0,0,24,32\\
&= 2;59,47,35,44 - 0;0,0,24,32\\
&= 2;59,47,11,12\\
b_1 &= b_0 + (0;0,4)d_1 + 2(0;0,4)d_1\\
&= 0;0,12 + (0;0,4)(0;2) + 2(0;0,4)(0;2)\\
&= 0;0,12 + 0;0,0,8 + 0;0,0,16\\
&= 0;0,12,24\\
g_1 &= g_0 + d_1 + 2d_1\\
&= 3 + 0;2 + 0;4\\
&= 3;2 + 0;4\\
&= 3;6\\
c_1 &= c_0 - (a_0 d_1 - b_0(d_1)^2 - (0;0,4)(d_1)^3)\\
&= 0;8,28,33,59,34,28,15 - ((2;59,48)(0;2) - (0;0,12)(0;2)^2 - (0;0,4)(0;2)^3))\\
&= 0;8,28,33,59,34,28,15 - (0;5,59,36 - 0;0,0,0,48 - 0;0,0,0,0,32)\\
&= 0;8,28,33,59,34,28,15 - 0;5,59,35,11,28\\
&= 0;2,28,58,48,6,28,15
\end{aligned}
$$

The value d_2 is chosen to be the largest number of seconds so that $d_2 \cdot a_1$ does not exceed c_1, so $d_2 = 0;0,49$. The numbers in the third step are shown in blue:

$$
\begin{aligned}
a_2 &= a_1 - d_2(b_1 + (0;0,4)d_2) - d_2(b_1 + 2(0;0,4)d_2)\\
&= 2;59,47,11,12 - (0;0,49)(0;0,12,24 + (0;0,4)(0;0,49)) - (0;0,49)(0;0,12,24 + 2(0;0,4)(0;0,49))\\
&= 2;59,47,11,12 - 0;0,0,10,10,16,4 - 0;0,0,10,12,56,8\\
&= 2;59,47,1,1,43,56 - 0;0,0,10,12,56,8\\
&= 2;59,46,50,48,47,48\\
b_2 &= b_1 + (0;0,4)d_2 + 2(0;0,4)d_2\\
&= 0;0,12,24 + (0;0,4)(0;0,49) + 2(0;0,4)(0;0,49)\\
&= 0;0,12,24 + 0;0,0,3,16 + 0;0,0,6,32\\
&= 0;0,12,33,48\\
g_2 &= g_1 + d_2 + 2d_2\\
&= 3;6 + 0;0,49 + 2 \cdot 0;0,49\\
&= 3;6 + 0;0,49 + 0;1,38\\
&= 3;8,27\\
c_2 &= c_1 - (a_1 d_1 - b_1(d_2)^2 - (0;0,4)(d_2)^3)\\
&= 0;2,28,58,48,6,28,15 - ((2;59,47,11,12)(0;0,49) - (0;0,12,24)(0;0,49)^2 - (0;0,4)(0;0,49)^3))\\
&= 0;2,28,58,48,6,28,15 - 0;2,26,49,32,8,48 - 0;0,0,0,8,16,12,24 - 0;0,0,0,0,2,10,43,16)\\
&= 0;2,28,58,48,6,28,15 - 0;2,26,49,23,50,24,52,44\\
&= 0;0,2,9,24,16,3,22,16
\end{aligned}
$$

The value d_3 is chosen to be the largest number of thirds so that $d_3 \cdot a_2$ does not exceed c_2, so $d_3 = 0;0,0,43$. The numbers in the third step are shown in green:

$$
\begin{aligned}
a_3 &= a_2 - d_3(b_2 + (0;0,4)d_3) - d_3(b_2 + 2(0;0,4)d_3) \\
&= 2;59,46,50,48,47,48 - (0;0,0,43)(0;0,12,33,48 + (0;0,4)(0;0,0,43)) \\
&\quad - (0;0,0,43)(0;0,12,33,48 + 2(0;0,4)(0;0,0,43)) \\
&= 2;59,46,50.48,47,48 - 0;0,0,0,9,0,16 - 0;0,0,0,9,0,17 \\
&= 2;59,46,50,39,47,32 - 0;0,0,0,9,0,17 \\
&= 2;59,46,50,30,47 \\
b_3 &= b_2 + (0;0,4)d_3 + 2(0;0,4)d_3 \\
&= 0;0,12,33,48 + (0;0,4)(0;0,0,43) + 2(0;0,4)(0;0,0,43) \\
&= 0;0,12,33,48 + 0;0,0,0,2,52 + 0;0,0,0,5,44 \\
&= 0;0,12,33,56,36 \\
g_3 &= g_2 + d_3 + 2d_3 \\
&= 3;8,27 + 0;0,0,43 + 2 \cdot 0;0,0,43 \\
&= 3;8,27 + 0;0,0,43 + 0;0,1,26 \\
&= 3;8,29,9 \\
c_3 &= c_2 - (a_2 d_3 - b_2(d_3)^2 - (0;0,4)(d_3)^3) \\
&= 0;0,2,9,24,16,3,22,16 - ((2;59,46,50,48,47,48)(0;0,0,43) \\
&\quad - (0;0,12,33,48)(0;0,0,43)^2 - (0;0,4)(0;0,0,43)^3)) \\
&= 0;0,2,9,24,16,3,22,16 - (0;2,8,50,34,24,58,15,24 \\
&\quad - 0;0,0,0,0,0,6,27,9,36,12 - 0;0,0,0,0,0,0,0,1,28,20,28) \\
&= 0;0,2,9,24,16,3,22,16 - 0;0,2,8,50,34,18,31,4,17 \\
&= 0;0,0,0,33,41,44,51,11,43
\end{aligned}
$$

Up to this point in the grid in figure 3.7, all calculations have been kept to full precision and are correct. During this fourth iteration, the compiler begins to round values as he terminates digits that extend beyond the 10th column in the 1st and 2nd sections of the grid, as seen in the following instances: 0;0,0,0,9,0,15,27,6 is rounded to 0;0,0,0,9,0,16 and 0;0,0,0,9,0,17,30,32 is rounded to 0;0,0,0,9,0,17 and 2;59,46,50,30,47,15 is rounded to 2;59,46,50,30,47. These do not reveal a consistent method of rounding on the part of our calculator. There is also a minor computational mistake in which 0;0,2,8,50,34,18,31,4,19,27,32 is rounded to 0;0,2,8,50,34,18,4,17.

The value d_4 is chosen to be the largest number of fourths so that $d_4 \cdot a_3$ does not exceed c_3, so $d_4 = 0;0,0,0,11$. The numbers in the fifth iteration are shown in violet:

$$\begin{aligned}
a_4 &= a_3 - d_4(b_3 + (0;0,4)d_4) - d_4(b_3 + 2(0;0,4)d_4) \\
&= 2;59,46,50,30,47 - (0;0,0,0,11)(0;0,12,33,56,36 + (0;0,4)(0;0,0,0,11)) \\
&\qquad - (0;0,0,0,11)(0;0,12,33,56,36 + 2(0;0,4)(0;0,0,0,11)) \\
&= 2;59,46,50,30,47 - 0;0,0,0,0,2 - 0;0,0,0,0,2 \\
&= 2;59,46,50,30,45 - 0;0,0,0,0,2 \\
&= 2;59,46,50,31 \\
b_4 &= b_3 + (0;0,4)d_4 + 2(0;0,4)d_4 \\
&= 0;0,12,33,56,36 + (0;0,4)(0;0,0,0,11) + 2(0;0,4)(0;0,0,0,11) \\
&= 0;0,12,33,56,36 + 0;0,0,0,0,0,44 + 0;0,0,0,0,1,28 \\
&= 0;0,12,33,56,38,12 \\
g_4 &= g_3 + d_4 + 2d_4 \\
&= 3;8,29,9 + 0;0,0,0,11 + 2 \cdot 0;0,0,0,11 \\
&= 3;8,29,9 + 0;0,0,0,11 + 0;0,0,0,22 \\
&= 3;8,29,9,33 \\
c_4 &= c_3 - (a_3 d_4 - b_3(d_4)^2 - (0;0,4)(d_4)^3) \\
&= 0;0,0,0,33,41,44,51,11,43 - ((2;59,46,50,30,47)(0;0,0,0,11) \\
&\qquad - (0;0,12,33,56,36)(0;0,0,0,11)^2 - (0;0,4)(0;0,0,0,11)^3)) \\
&= 0;0,0,0,33,41,44,51,11,43 - (0;0,0,0,32,57,35,15,38,37 \\
&\qquad - 0;0,0,0,0,0,0,0,0,25,20,27,8,36 - 0;0,0,0,0,0,0,0,0,0,0,0,1,28,44) \\
&= 0;0,0,0,33,41,44,51,11,43 - 0;0,0,0,32,57,35,15,38,11,39,32,49,55,16 \\
&= 0;0,0,0,0,44,9,35,33,49
\end{aligned}$$

In the calculation for a_4, the writer subtracts 0;0,0,0,0,2 from a_3 but does not bother to set down the subtrahend. Since he rounds a_4 to only four fractional sexagesimal places, he has no need to subtract the other 0;0,0,0,0,2. As the values of the computed quantities get smaller, the calculator generally rounds them to fewer sexagesimal places, most notably in the following instances: 0;0,0,0,0,2,18,13,22,44,4 is rounded to 0;0,0,0,0,2 and 0;0,0,0,0,2,18,13,22,52,8 is rounded to 0;0,0,0,0,2. Moreover, the sexagesimal digit 49 in the last place of the value $c_4 = 0;0,0,0,0,44,9,35,33,49$ is apparently a computational error for 27.

The value d_5 is chosen to be the largest number of fifths so that $d_5 \cdot a_4$ does not exceed c_4, so $d_5 = 0;0,0,0,0,14$. For the sixth iteration with $d_5 = 0;0,0,0,0,14$ the compiler only calculates c_6, shown below in orange:

$$\begin{aligned}c_5 &= c_4 - (a_4 d_5 - b_4(d_5)^2 - (0;0,4)(d_5)^3)\\ &= 0;0,0,0,0,44,9,35,33,49 - ((2;59,46,50,31)(0;0,0,0,0,14)\\ &\quad - (0;0,12,33,56,38,12)(0;0,0,0,0,14)^2 - (0;0,4)(0;0,0,0,0,14)^3))\\ &= 0;0,0,0,0,44,9,35,33,49 - (0;0,0,0,0,41,56,55,47,14\\ &\quad - 0;0,0,0,0,0,0,0,0,0,0,41,2,53,0,47,12 - 0;0,0,0,0,0,0,0,0,0,0,0,0,0,0,3,2,56)\\ &= 0;0,0,0,0,44,9,35,33,49 - 0;0,0,0,0,41,56,55,47,13,59,18,57,6,59,9,45,4\\ &= 0;0,0,0,0,2,12,39,46,35\end{aligned}$$

In this last calculation, two very small numbers are to be subtracted from 0;0,0,0,0,41,56,55,47,14, but the two subtracted numbers are so small that they would round to zero. In theory, then, the final quantity c_5 was achieved by rounding 0;0,0,0,0,2,12,39,46,35,0,41,2,53,0,50,14,56 to 0;0,0,0,0,2,12,39,46,35. However, it is likely that the compiler simply ignored the two small subtrahends.

	1			2			49			43			11			14
1st	3	8	24	33	59	34	28	15	0	0	0	0	0	0	0	0
	2	59	56													
		8	28													
		5	59	35	11	28										
		2	28	58	48	6										
		2	26	49	23	50	24	52	44							
			2	9	24	16	3	22	16							
			2	8	50	34	18	31	4	17						
					33	41	44	51	11	43						
					32	57	35	15	38	15						
						44	9	35	33	49						
						41	56	55	47	14						
						2	12	39	46	35						
2nd	3	0	0													
			4													
	2	59	56													
			8													
		2	59	48												
					24	16										
		2	59	47	35	44										
					24	32										
			2	59	47	11	12									
						10	10	16	4							
			2	59	47	1	1	43	56							
						10	12	56	8							
				2	59	46	50	48	47	48						
								9	0	16						
				2	59	46	50	39	47	32						
								9	0	17						
					2	59	46	50	30	47						
					2	59	46	50	30	45						
						2	59	46	50	31						
3rd	1		3	2	3	6	49									
	2			4	3	7	38									
							3	8	27	43						
									28	26						
									3	8	29	9	11			
													22			
											3	8	29	9	33	14

Figure 3.7: Translated version of the computational grid, shown in facsimile in figure 3.8 and transcribed in table 2.1, for the method attributed to Dayānat Khān.

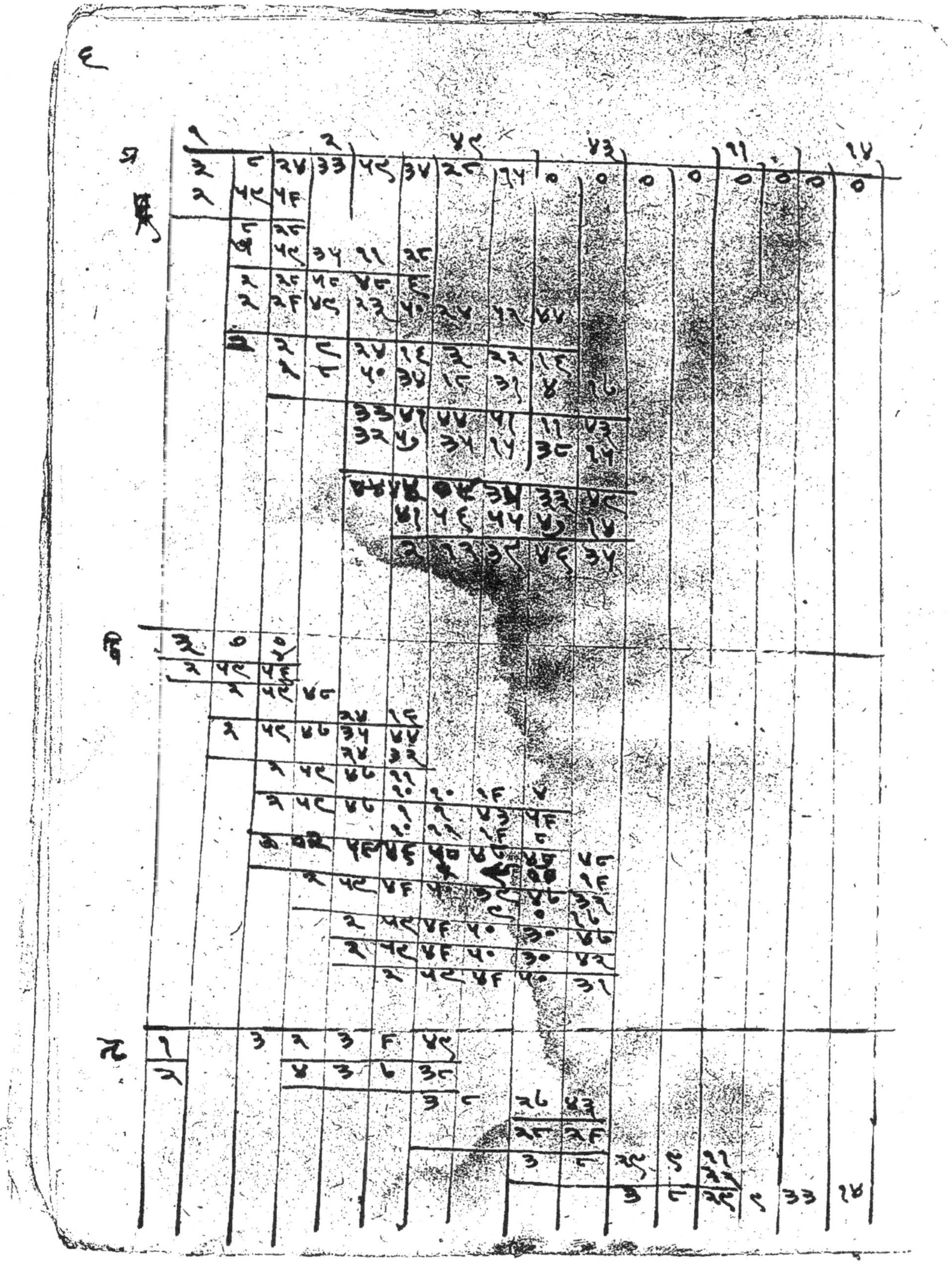

Figure 3.8: Excerpt from a manuscript page of the second part (J_2) of the *Jyācāpopapatti* (Jaipur MS 269:I.13 i 3, f. 6v), showing Dayānat Khān's successive-digits iteration worked calculation grid. This grid is transcribed in table 2.1 and translated in figure 3.7.

Chapter 4
Al-Bīrjandī and His Predecessors on Approximating the Sine of 1°

4.1 The Early Textual Tradition of al-Kāshī's Sine 1° Methods: Ulugh Beg and al-Bīrjandī

One of the earliest surviving expositions of al-Kāshī's iterative sine computation discoveries is an Arabic treatise of uncertain authorship, composed at the Samarqand observatory of the Timurid astronomer-prince Mīrzā Ulugh Beg. The various identifications of this text in its known manuscripts include the following descriptions (Rosenfeld and Hogendijk 2003, p. 32):

> A treatise on the determination of the sine of one degree according to operations established by arithmetic and geometric rules by the method of Ghiyāth al-Dīn al-Kāshī.
>
> A treatise on the determination of the sine of one degree with true precision, determined by the most perfect of the geometers, Jamshīd al-Qāsānī, edited and revised in this letter by Qāḍī-Zādeh al-Rūmī, the author of the commentary on the Chaghmīnī.

Alternatively, it has been argued (Rosenfeld and Hogendijk 2003, pp. 27–31) that the author was not in fact the mathematican al-Rūmī, a known colleague of al-Kāshī at Samarqand, but Ulugh Beg himself.

This work, henceforth designated "the Arabic treatise", appears to be closely related to the Persian commentary *Sharḥ-i Zīj-i Jadīd-i Sulṭānī* on the *Zīj-i Ulugh Beg*, composed ca. 1523 by the ᶜAbd al-ᶜAlī al-Bīrjandī briefly alluded to in section 1.1.2. Al-Bīrjandī's teacher was a different al-Kāshī, Manṣūr ibn Muᶜīn al-Dīn, who also worked as an astronomer at the Samarqand observatory. (He was in fact the "Muᶜīn al-Dīn" whose notes from a manuscript of the *Zīj-i Khāqānī* were quoted in section 1.1.2.) Al-Bīrjandī's commentary is evidently the work described at the start of MS J_2 as the *vyākhyā* (commentary) "*Śarahavirjaṃdī* of the *Ulugvegī zīj*".

Ulugh Beg in his introduction to that *zīj* states that its computations were initially undertaken by (Ghiyāth al-Dīn Jamshīd) al-Kāshī and the abovementioned al-Rūmī; after their deaths, their labors were continued by another Samarqand astronomer, ᶜAlī al-Qūshjī (d. 1474) (Sédillot 1839, pp. 289–290), (Sédillot 1853, p. 5). As al-Kāshī died in 1429, al-Rūmī after 1440, and Ulugh Beg himself in 1449, work on the *zīj* must have commenced in the 1420s and continued into the 1440s. Its discussion of the trigonometry required for spherical astronomy is contained in the second chapter of the introductory section preceding the main matter of the *zīj* tables.

Section 4.2 edits and translates material in al-Bīrjandī's *Sharḥ* at the end of the *zīj*'s trigonometry chapter that overlaps with our *jyācāpa* text. We have consulted somewhat low-resolution digital copies of excerpts from three manuscripts in the Majlis Library in Tehran, with shelfmark identifications 4716, 6003, and Senate 840, as specified rather sketchily in the following table. Lacking fuller information on the manuscripts, we have not ventured on any more detailed paleographic description of their characteristics or content (nor have we indicated their page-break locations in the edited Persian text in section 4.2).

Siglum	Start of excerpt	End of excerpt	Notes
T4716	f.32a, line 16	f.34b, line 8	Roughly written foliation in upper left corner of side **a** of leaves; rubricates diagram and table lines as well as some abjad numbers.
T6003	PDF page 30, 2nd opening, side **a**, line 9	PDF page 32, 1st opening, side **a**, line 19	No visible foliation; PDF images show two openings per page; rubricates diagram and table lines, abjad numbers.
TS840	PDF page 95, side **b**, line 9	PDF page 102, side **b**, line 6	No visible foliation; one opening per PDF page; omits diagrams and abjad numbers; no rubrication except lines of (empty) table.

As in our Sanskrit editions and translations in chapter 2, paragraph breaks have been added for readability. In section 4.3 we analyze the mathematical content of the excerpt, with reference to the abovementioned Arabic treatise as presented in

C. Montelle et al., *From Samarqand to Jaipur*, Sources and Studies in the History of Mathematics and Physical Sciences, https://doi.org/10.1007/978-3-031-89087-1_4

(Rosenfeld and Hogendijk 2003), as well as to the abridged French translation of Ulugh Beg's *zīj*—especially the excerpts from Mīram Čelebī's 1499 commentary pertaining to sin 1° approximations—in (Sédillot 1853, pp. 69–83).

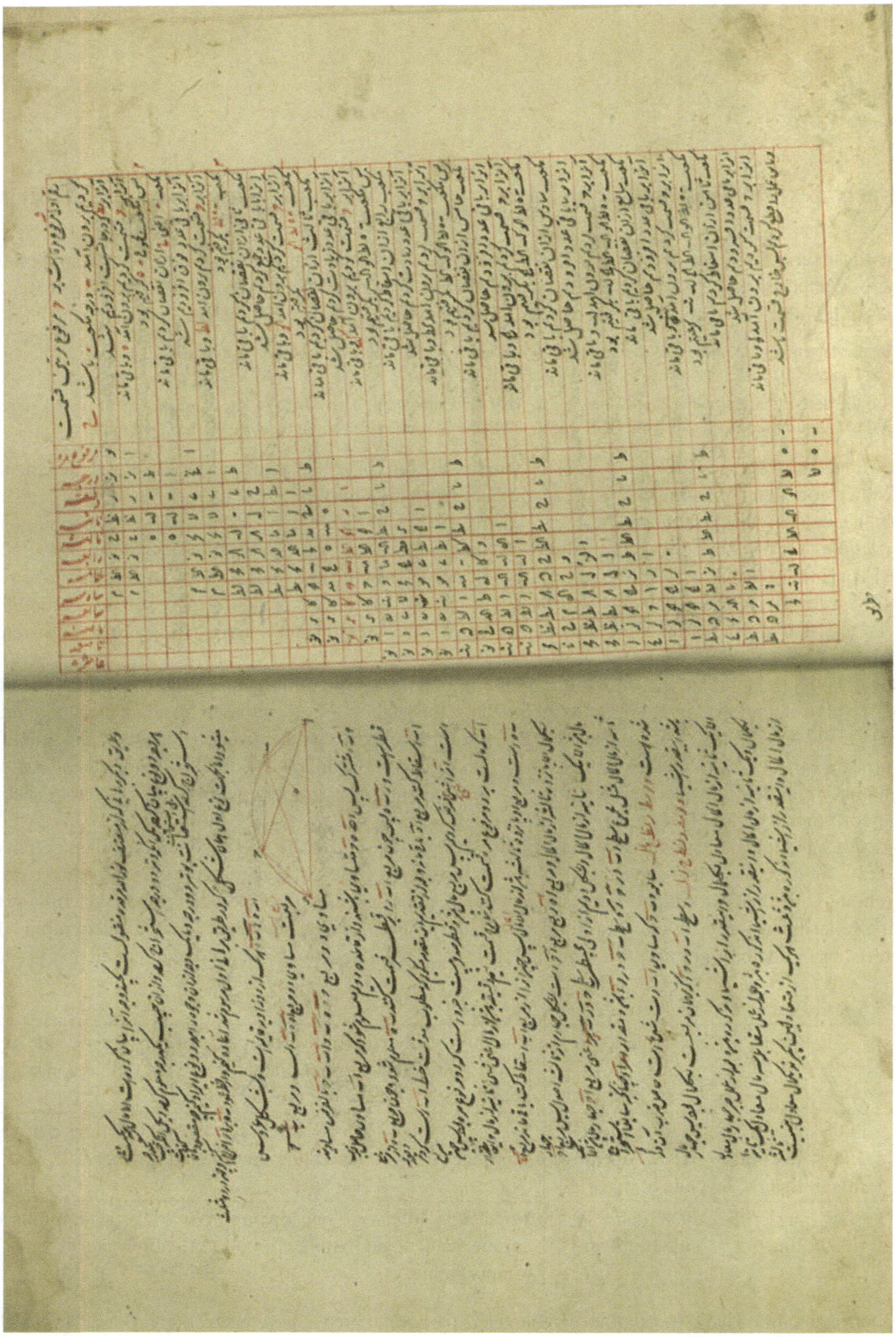

Figure 4.1: Excerpt from a manuscript of al-Bīrjandī's commentary on the *Zīj-i Ulugh Beg* (Tehran Majlis MS 6003, PDF p. 31), showing text and a computation grid.

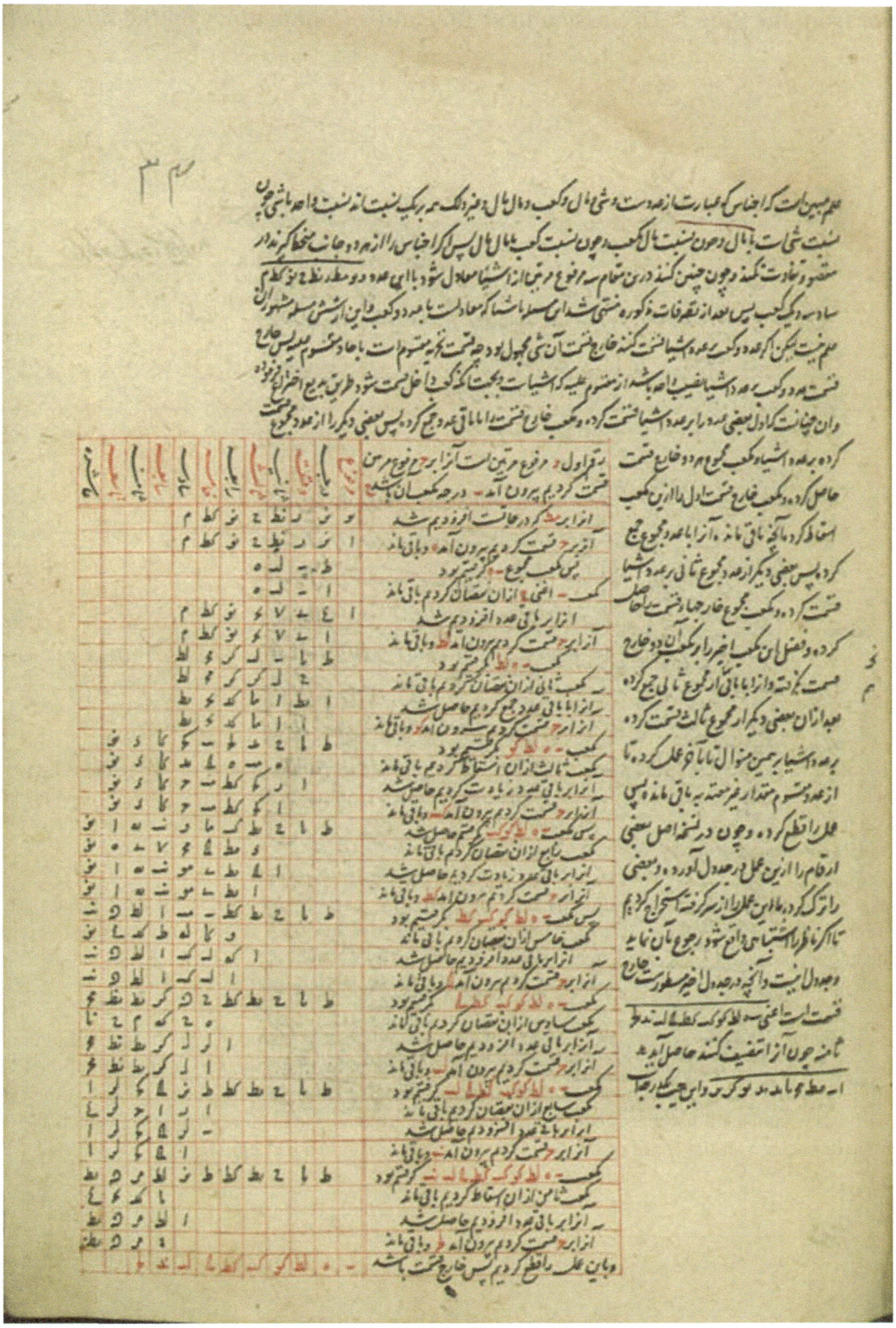

Figure 4.2: Excerpt from a manuscript of al-Bīrjandī's commentary on the *Zīj-i Ulugh Beg* (Tehran Majlis MS 4716, f. 33a), showing text and a computation grid.

4.2 Excerpt from the Sine 1° Discussion in al-Bīrjandī's Commentary on the *Zīj-i Ulugh Beg*

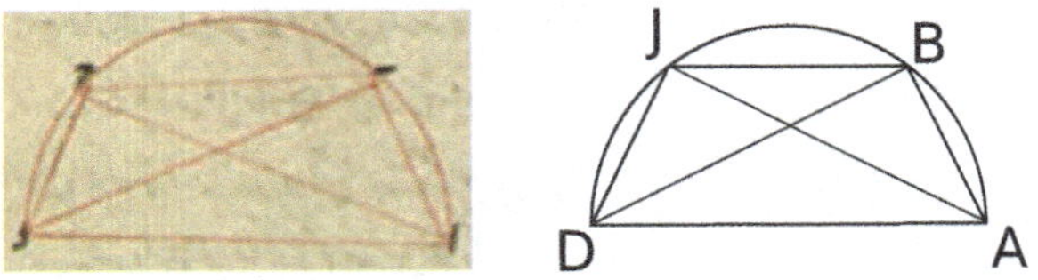

Figure 4.3: Diagram on f.32a of MS Tehran Majlis 4716 (facsimile and reconstruction)

و چون طریقة استخراج جیب یکدرجة بتقریب معلوم شد طریق استخراج آن برهان نیز ایراد کنیم.[1] و آن دو طریقست[2] یکی آنکه سلطان المهندسین غیاث الدین جمشید استخراج نموده[3] و دیگر آنکه سلطان شهید سعید[4] مضف بیان فرموده. و ما این هر دو بر سبیل اختصار ایراد کنیم.[5]

اما بجهت طریق اول فرض کنیم که قوس ابجد شش درجة است و انرا سه قسم متساوی کنیم بر[6] نقطهای ب ج و اوتار اب اج اد بج بد جد وصل کنیم. و از جیب پانزده درجة و جیب هژده[7] درجة که سابقا معلوم شد جیب سه درجه استخراج کردیم بود ج ح کد لج نط لد کح ید ن ثامنة. پس آنرا مضاعف ساختیم[8] حاصل شد وتر اد و یو مط ز نط ح نو کط م.

و مطلوب معرفت وتر اب است که دو درجة است. پس میگویم در شکل دوم از مقالة اول مجسطی بین شده است که هر ذواربعة[9] که در دائرة واقع شود مجموع دو سطح هر ضلع در مقابل او متساوی سطح یکی از دو قطراوست در دیگری.[10] و[11] از شکل چهاروم بهان مقالة مستفادی شود که مربع وتر نصف قوسی متساوی سطح نصف قطرست در فضل قطر بر وتر تمام اصل آن قوس تا نصف دور.

And since the method of extracting the sine of one degree became approximately known, we demonstrate the method of extracting that [by] proof also. And that is two methods: one which the Sultan of Geometers Ghiyāth al-Dīn Jamshīd [al-Kāshī] derived, and the other [for] which the martyred Sultan [Ulugh Beg], the blessed ruler, prescribed the explanation. And we demonstrate this for each of the two by way of a summary.

So for the sake of the first method, we assume that arc $ABJD$ is six degrees, and we make it three equal parts at points B, J, and we join the chords AB, AJ, AD, BJ, BD, JD. And from the sine of fifteen degrees and the sine of eighteen degrees, which became known previously, we calculate the sine of three degrees, [which] is 3; 8, 24, 33, 59, 34, 28, 14, 50, [to] eighths. Then we double that; the result becomes the chord AD, 6; 16, 49, 7, 59, 8, 56, 29, 40.

[What is] desired is the acquisition of the chord of AB, which is two degrees. Then we say in the second figure from the first chapter of the *Almagest* [it] is demonstrated that [for] every quadrilateral that falls in a circle, the sum of the two products of each side with its opposite is equal to the product of one of its two diagonals with the other. And from the fourth figure in that chapter [it] was established that the square of the chord of half the arc is equal to the product of half the diameter with the excess of the diameter over the chord of the supplement of that arc up to a half circle.

1 آن نیز برهان معلوم کنیم TS840
2 طریق است T6003
3 کرده T4716
4 شهید ایراد سعید TS840
5 اختصار بیان کنیم T6003
6 کنیم و بر T6003
7 درجه و هژده T4716, درجة جیب هیژده T6003
8 مضاعف کردیم T6003
9 که ذواربعة T6003
10 سطح یکی از اول مجسطی بین شده است که هر ذواربعة که در دائرة واقع شود مجموع دو سطح هر ضلع در مقابل الی دو قطراوست در دیگری TS840
11 و om. T6003

و ظاهر است که اب جد متساویند و همچنین اج بد پس حاصل[12] ضرب اب در جد مربع باشد که در اصطلاح اهل جبر و مقابلة انرا مآل گویند و حاصل ضرب بج که مجهولست در اد باشد این قدر از اشیاء و یو مط ز نط ح نو کط م. و مجموع این دو حاصل متساوی مربع اج است یعنی سطح اج در بد بحکم مقدمة اولی. و مربع اب متساوی سطح بصف قطرست که شصت است در فضل قطر بر وتر تمام اج تا نصف دور بحکم مقدمة ثانیة چه قوس اج ضعف قوس اب است.

پس چون مربع اب که مال است بر شصت قسمت کند[13] خارج قسمت جزوی باشد از شصت جزو از مال و این فضل قطرست بر وتر تمام اج. پس چون آنرا از قطر که صد و بیست جزوست نقصان کنند باقی ماند وتر تمام اج صد و بیست جزو الا یک جزو از شصت جزو از مال.[14]

و در علم جبر و مقابلة مقرر شده است که چون مربع عددی خواهند که در ان استثنا باشد اول مستثنی منه را در نفس او ضرب کنند پس مستثنی را در نفس او ضرب کنند و هر دو را جمع کنند وضعف حاصل ضرب مستثنی را در مستثنی منه از مجموع مذکور استثنا کنند.

پس بجهت معرفت وتر تمام اج اول مربع صد و بیست یک قسم بود ۱۴۴۰۰ و چون آنرا مرفوع کنند چهار مرفوع مرة باشد که چهار مال است. و مربع جزوی از شصت جزو از مال جزوی باشد از سه هزار و ششصد جزو از مال مال زیرا که حاصل ضرب جزو مال در نفس او جزو مال المال است.[15] و ضعف حاصل ضرب صد و بیست در جزوی از شصت جزو مال[16] چهار مال است. پس مربع وتر تمام اج چهار مرفوع مرة باشد از درجات و جزوی[17] از سه هزار و ششصد جزو از مال مال الا چهار مال.

And it is clear that AB, JD are equal, and in the same way AJ, BD; so the product of multiplying AB by JD is a square, which in the terminology of algebraists they call *māl*. And the product of multiplying BJ, which is unknown, by AD is this quantity of the *shay* [pl.]: 6; 16, 49, 7, 59, 8, 56, 29, 40. The sum of these two products is equal to the square of AJ, i.e., the product of AJ with BD, on account of the first theorem [Ptolemy's theorem]. And the square of AB is equal to the product of half the diameter, which is sixty, with the excess of the diameter over the chord of the supplement of AJ up to a half-circle, on account of the second theorem [*Almagest* 1.4], since arc AJ is double arc AB.

So when the square of AB, which is the *māl*, is divided by sixty, the quotient is a sixtieth part of the *māl*, and this is the excess of the diameter over the chord of the supplement of AJ. So when they subtract that from the diameter which is a hundred and twenty parts, [what] remains [is] the chord of the supplement of AJ, a hundred and twenty parts less one sixtieth part of the *māl*.

In the science of algebra it is established that if they intend [to find] the square of some number in which there is a subtraction, first they multiply the minuend by itself, then they multiply the subtrahend by itself, and they add each of the two; and they subtract twice the product of multiplying the subtrahend by the minuend from the [above-]mentioned sum.

So for the sake of obtaining the chord of the supplement of AJ, first the square of one hundred and twenty in one [integer] unit is 14400; and when they raise it [sexagesimally], four is raised [one] turn, which is four *māl*. And the square of a sixtieth part of the *māl* is a three-thousand-six-hundredth part of the *māl-māl*, because the product of multiplying parts of the *māl* by itself is parts of the *māl* of the *māl*. Twice the product of multiplying a hundred and twenty by a sixtieth part of the *māl* is four *māl*. So the square of the chord of the supplement of AJ is four raised [one] turn from degrees and a three-thousand-six-hundredth part of the *māl-māl*, less four *māl*.

---

12 حاصل om. TS840

13 قسمت کنند T6003, T4716

14 صد و بیست جزو بر وتر الا یک جزو از شصت جزو از مال T6003, صد و بیست جزو از مال الا یک جزو از شصت جزو از مال TS840

15 حاصل ضرب جزو مال المالست T6003, حاصل ضرب جزو مال در نفس خودح مال المال است TS840

16 جدوو مال T6003

17 مرقوع باشد از درجات جزوی T6003, مرقوع باشد از درجات و جزوی TS840

و چون مربع وتر قوسی از مربع قطر اسقاط کنند مربع وتر تمام آن قوس تا نصف دور بماند چه از قطر و این دو وتر مثلث قائم الزاویة حاصل شود که وتر آن زاویة قائمة قطر بود[18] زیرا که زاویة نصف دائرة قائمة است چنانچه مبین شده است در[19] ثالثة اصول. و مربع وتر قائمة مساوی مجموع دو مربع دو ضلع اوست بشکل عروس.

و هم در علم جبر و مقابلة مقرر شده است که چون در عدد استثنا باشد وخواهند[20] که آنرا از عدد دیگر نقصان کنند اول مستثنی را[21] خدف کنند و مثل مستثنی بر منقوص منه زیادت کنتد پس آن عددرا نقصان کنند.[22] پس میگویم مربع قطر چهار مال است پس چهار مال مستثنی است از منقوص نقصان کردیم و بر منقوص منه که[23] چهار مال است[24] اضافة کردیم پس منقوص منه هشت مال شد.[25] پس چون مربع وتر تمام اج از ان اسقاط کردیم باقی ماند مربع وتر اج چهار مال الا جزوی از سه هزار و ششصد جزو از مال مال.

و سابقا گفتیم که مربع وتر اج مساوی سطح اب در جد است که مال است وسطح بج در اد که آن هست از اشیا این قدر و یو مط ز نط ح نو کط م. پس چهار مال الا جزوی از سه هزار و ششصد جزو از مال مال معادل باشد با مال[26] و این قدر اشیا که مذکور شد.

و چون مستثنی از معادل اول اسقاط کنند و مثل آن[27] بر معادل ثانی اضافة کنند چهار مال معادل شود با یک مال[28] و این قدر اشیا و یو مط ز نط ح نو کط م و جزوی از سه هزار و ششصد جزو از مال مال و این عمل را جبر گویند.

And when they subtract the square of the chord of an arc from the square of the diameter, the square of the chord of the supplement of that arc up to half a circle remains, since from the diameter and these two chords a right-angled triangle was produced [so] that the chord of that right angle was the diameter, because the angle of half a circle is right, just as is demonstrated in the third [book] of the *Elements*. And the square of the chord of the right [angle] is equal to the sum of the two squares of its two sides, by the "Bride Figure".

And also in the science of algebra it is established that when in a number there is a subtraction and they intend [to] subtract that from another number, first they cut out [?] the subtrahend and they add the equivalent of the subtrahend to the minuend; then they subtract that number. So I say the square of the diameter is four *māl*, so we subtract four *māl* [which] is negative [?] from the subtrahend, and we add to the minuend which is four *māl*, so the minuend is eight *māl*. So when we subtract the square of the chord of the supplement of *AJ* from that, [there] remains the chord of *AJ*, four *māl* less a three-thousand-six-hundredth part of the *māl-māl*.

And previously we said that the square of the chord of *AJ* is equal to the product of *AB* with *JD*, which is the *māl*, and the product of *BJ* with *AD*, which is this amount of the *shay*: 6; 16, 49, 7, 59, 8, 56, 29, 40. So four *māl* less a three-thousand-six-hundredth part of the *māl-māl* is equal to [one] *māl* and this amount of the *shay* which was mentioned.

And when they subtract the negative from the first equal, and they add the equivalent of that to the second equal, four *māl* is equal to one *māl* and this amount of the *shay*, 6; 16, 49, 7, 59, 8, 56, 29, 40, and a three-thousand-six-hundredth part of the *māl-māl*; and they call this operation "restoration" (*jabr*).

---

18 وتراو زاویة قائمة مط بود T6003

19 مبین شد در T6003, مبین شده در TS840

20 کواهند TS840

21 کنند و مستثنی را T6003

22 پس آن عددرا نقصان کنند om. TS840

23 که om. TS840

24 مالست T4716

25 باشد TS840

26 مالی T6003, T4716

27 و مثل او T6003

28 یکیمال T6003, یکی مال TS840

و چون آنچه مشترکست[29] در متعادلین خدف کنند بماند سه مال معادل این قدر[30] اشیا و یو مط ز نط ح نو کط م ثامنة و جزوی از سه هزار و ششصد جزو از مال مال و این عمل را مقابلة گویند. و هم از قواعد[31] آن علم است[32] که چون در یکی از متعادلین کسری باشد آن کسر را تمام سازند و مثل آن[33] بر معادل دیگر اضافة کنند و این عمل را تکمیل گویند.

پس[34] چون تکمیل کسر کنند درین مقام ده هزار و هشتصد مال وآن سه مرفوع مرتین بود از مال معادل شود باین قدر اشیا و یو مط ز نط ح نو کط م سادسه و یک مال مال.[35] زیرا که چون سه را در سه هزار و ششصد ضرب کنند ده هزار و هشتصد شود وچون ارقام[36] این اعداد در سه هزار وششصد ضرب کنند هر رقم بدو مرتبة رفع شود. مثلا سه هزار و ششصد ثامنة شصت سابعة است که یک سادسة باشد وعلی هذا القیاس.[37]

و در همان علم مبین است که اجناس که عبارت[38] از عدد است[39] و شی و مال وکعب و مال مال و غیر ذلک ثمه[40] بر یک نسبت اند. و نسبت[41] واحد با شی چون نسبت شی است با مال وچون نسبت مال با کعب و چون نسبت کعب با مال مال. پس اگر اجناس را از هر دو جانب میخط[42] گیرند در مقصود تفاوت نکند. و چون چنین کنند در ین مقام[43] سه مرفوع مرتین از اشیا معدل شود با این عدد و یو مط ز نط ح نو کط م سادسة و یک کعب.

And since whatever is common in the two equals they cut out, [there] remain three *māl* equal to this amount of the *shay*, 6; 16, 49, 7, 59, 8, 56, 29, 40, [to] eighths, and a three-thousand-six-hundredth part of the *māl-māl*; and they call this operation "confrontation" (*muqābala*). And also it is [known] from the rules of that science that when in one of two equals there is a fraction, they make that fraction the complement and they add its equivalent to the other equal; and they call this operation "completion" (*takmīl*).

So when they make the completion of the fraction—in this place ten thousand eight hundred *māl*, and that is three raised two turns [as the coefficient] of the *māl*—the equal [other side] has this amount of the *shay*, 6, 16, 49; 7, 59, 8, 56, 29, 40, [to] sixths, and one *māl-māl*. Because when they multiply three by three thousand six hundred, ten thousand eight hundred are [the result], and when they multiply the places of these numbers by three thousand six hundred, each place is raised by two degrees. For instance, three thousand six hundred [to] eighths is sixty [to] sevenths, which is one [to] sixths, and so on.

And in that selfsame science it is clear that [variable] "kinds" (which is a technical term from computation, [meaning] *shay* and *māl* and cube and *māl-māl* and [others] besides this) therefore are in the same ratio. And the ratio of the unit to *shay* [is] as the ratio of *shay* is to *māl* and as the ratio of *māl* to cube and as the ratio of cube to *māl-māl*. So if they write the kinds on each of the two sides, [there] will not be a contradiction in [what] is intended. And when they do [it] in this way, in this place three raised two turns from the *shay* is equal to this number, 6, 16, 49; 7, 59, 8, 56, 29, 40, [to] sixths, and one cube.

29 مشترک است TS840
30 سه مال ح معادل ایتقدر T6003, سه مال معادل و این قدر T4716
31 و هم ز قوعد T6003
32 عامست T4716
33 مثل او T4716
34 پس om. TS840
35 سادسه یکمال مال T6003
36 مقام TS840
37 هذ القیاس TS840
38 که اخباس عبارت TS840
39 عددست T4716
40 [?] T6003
41 اند نسبت T4716
42 منخط T4716
43 نکند در ینمقام T6003

پس بعد[44] از تصرفات مذکورة منتها شد این مسلة باشیا[45] که معادل است[46] با عدد و کعب و این از شش مسلة مشهور ان علم نیست. لیکن اگر عدد و کعب بر عدد اشیا قسمت کنند خارج قسمت آن شی مجهول بود چه قسمت تجربه[47] مقسوم است باحاد مقسوم علیه. پس خارج قسمت عدد وکعب بر عدد اشیا نصب واحد باشد از مقسوم علیه که اشیا است. و بجهت انکه کعب داخل قسمت شود طریق بریع اختراج فرموده.

و ان چنان است که اول بعض عدد را بر عدد اشیا قسمت کرده و مکعب خارج قسمت را با باقی عدد جمع کرده. پس بعض دیگر[48] را از عدد مجموع قسمت کرده بر عدد اشیا ومکعب مجموع هر دو خارج قسمت حاصل کرده ومکعب خارج قسمت اول را ازین مکعب اسقاط کرده آنچه باقی ماند آنرا با عدد مجموع جمع کرده.

پس بعض دیگر از عدد مجموع ثانی بر عدد اشیا قسمت کرده ومکعب[49] مجموع هر دو خارج قسمت را[50] حاصل کرده. و فضل این مکعب اخر را بر مکعب آن دو[51] خارج قسمت بگرفت و انرا با باقی از مجموع ثانی جمع کرده. بعد از ان[52] بعض دیگر از مجموع ثالث قسمت کرده بر عدد اشیا بر همین منوال تا بآخر عمل کرده تا از عدد مقسوم مقدار غیر معتد به باقی ماند پس عمل را قطع کرده.

So after the conclusion of the [above-]mentioned operations, this problem [is] about a *shay* which is equal to a number and a cube; and this is not among the six well-known problems of this science. But if they divide the number and the cube by the number of the *shay*, the quotient, that *shay*, is unknown, since division is the assaying of the dividend by the units of the divisor. So the quotient of the number and the cube [divided] by the number of the *shay* is one power of the divisor, which is the *shay*. And for the sake of [the case] where the cube is involved [in] division, a method by increasing extraction [is] prescribed.

And that is in such a manner: First, dividing some [part] of the number by the number of the *shay*, and adding the cube of the quotient to the remaining number. Then dividing some other [part] from the number of the sum by the number of the *shay*, and bringing out the cube of the sum of each of the two quotients, and subtracting the cube of the first quotient from this cube, adding what remains to the number of the sum.

Then dividing some other [part] from the number of the second sum by the number of the *shay*, and bringing out the cube of the sum of each of the two quotients. Then taking the excess of that last cube over the cube of those two quotients, and adding that to the remainder from the second sum. After that, dividing some other [part] from the third sum by the number of the *shay*, in this same manner until ending the operation—until the amount of the number of the dividend [is] not counted in the remainder, then terminating the operation.

---

44 بعد om. T6003
45 با اشیا T6003
46 معادلست T4716
47 بتجربه T6003
48 TS840 [begin lacuna]
49 بر عدد اشیا قسمت کرده بر عدد اشیا مکعب T6003
50 مجموع خارجهار قسمت را T4716
51 مکعب او دو T6003
52 و بعد از ان T6003

و چون در نسخة اهل بعض ارقام را ازين عمل[53] در جدول آورده وبعض را[54] ترک کرده ما اين عمل را[55] از سر گرفته استخراج کرديم تا اگر ناظر را اشتباهی واقع شود رجوع بآن نمايد.[56] وجدول در صفحة ايندة است[57] وآنچه در جدول آخر مسطور است[58] خارج قسمت است[59] اعنى ب ه لط كو كب كط كح لب نب لج ثامنة. چون آنرا تنصيف كنند حاصل شود[60] ا ب مط مج يا يد مد يو كو يو ل[61] و اين جيب يكدرجة[62] است.

And because in copying, people [are] bringing in some figures from this operation in a table and leaving some out, we lay out this operation from the beginning until, if the reader is doubtful, repetition testifies to that. And the table is on the next page, and whatever is written in the end of the table is the quotient: I mean, 2; 5, 39, 26, 22, 29, 28, 32, 52, 33, [to] eighths [correctly, ninths]. When they halve that, the result is 1; 2, 49, 43, 11, 14, 44, 16, 26, 16, 30, and this is the sine of one degree.

53 ازين امال T6003
54 را om. T6003
55 و ما اين عمل T6003
56 نمايند T6003
57 وجدول اينست T4716
58 اخير مسطورست T4716
59 است om. T6003
60 حاصل ايرعد T4716
61 ل om. T4716
62 جيب درجة T6003

رقم اول **و** مرفوع مرتین است آنرا بر[63]**ج** مرفوع مرتین قسمت کردیم بیرون آمد **ب** درجة مکعب ان[64]باشد **ح**	درجة	درجة	دقيقة	ثانية	ثالثة	رابعة	خامسة	سادسة	سابعة	ثامنة	تاسعة	عاشرة
آنرا بر **مط** که درجاتست افزودیم شد	یو	نز	ز	نط	ح	نو	کط	م				
آنرا بر **ج** قسمت کردیم بیرون آمد **ه** وباقی ماند	ا	نز	ز	نط	ح	نو	کط	م				
پس مکعب مجموع **ب ه** کردیم بود		ط	ب	لب	ه							
مکعب **ب** اعنی **ح** ازان نقصان کردیم باقی ماند		ا	ب	لب	ه							
آنرا بر باقی عدد فوق[65]افزودیم شد	ا	نح	ی	لا	یج	نو	کط	م				
آنرا بر **ج** قسمت کردیم بیرون آمد **لط** وباقی ماند		ا	ی	لا	یج	نو	کط	م				
مکعب **ب ه لط** بگرفتیم بود		ط	یا	ب	لب	کز	مج	لط				
مکعب ثانی ازان نقصان کردیم باقی ماند			ح	ل	کز	کز	مج	لط				
آنرا با باقی عدد جمع کردیم حاصل شد		ا	یط	ا	ما	کد	یج	یط				
آنرا بر **ج** قسمت کردیم بیرون آمد **کو** وباقی ماند			ا	ا	ما	کد	یج	یط				
مکعب **ب ه لط کو** بگرفتیم بود		ط	یا	ح	ید	لج	یب	کج	کا	ک	نو	
مکعب ثالث ازان نقصان[66]کردیم باقی ماند				ه	مب	ه	کح	مد	کا	ک	نو	
آنرا بر باقی عدد زیادت کردیم حاصل شد			ا	ز	کج	کط	مب	ج	کا	ک	نو	
آنرا بر **ج** قسمت کردیم بیرون آمد **کب** وباقی ماند				ا	کج	کط	مب	ج	کا	ک	نو	
پس مکعب **ب ه لط کو کب** بگرفتیم بود[67]		ط	یا	ح	یط	کب	ما	و	نب	یه	۱	نو
مکعب رابع ازان نقصان[68]کردیم باقی ماند					ک	مط	کح	مج	لا	ی	ه	نو
آنرا بر باقی عدد زیادت کردیم حاصل شد				ا	کح	یط	ی	مو	نب	یه	ا	نو
آنرا بر **ج** قسمت کردیم بیرون آمد **کط** وباقی ماند					ا	یط	ی	مو	نب	یه	ا	نو
پس مکعب **ب ه لط کو کب کط** بگرفتیم بود		ط	یا	ح	یط	کط	ب	مب	ا	لط	ن	نب
مکعب خامس ازان نقصان کردیم باقی ماند						و	کا	له	ط	کد	مح	نو
آنرا بر باقی عدد افزودیم حاصل شد					ا	که	لب	کب	ا	لط	ن	نب
آنرا بر **ج** قسمت کردیم بیرون آمد **کح** وباقی ماند						ا	لب	کب	ا	لط	ن	نب
مکعب **ب ه لط کو کب کط کح** بگرفتیم بود		ط	یا	ح	یط	کط	ح	ن	کز	یط	نط	مج
مکعب سادس ازان نقصان کردیم باقی ماند							و[69]	ح	که	م	ح	نا
آنرا بر با باقی عدد افزودیم حاصل شد						ا	لز	ل	کز	یط	نط	مج
آنرا بر **ج** قسمت کردیم بیرون آمد **لب** وباقی ماند							ب[70]	ل	کز	نط	نط	مج
مکعب **ب ه لط کو کب کط کح لب** بگرفتیم بود		ط	یا	ح	یط	کط	ح[71]	نز	کح	کج	لز	ا
مکعب سابع ازان نقصان کردیم باقی ماند							ء[72]	ز	ا	ج	لز	یح
آنرا بر باقی عدد افزودیم حاصل شد							ب	لز	کح	کج	لز	ا
آنرا بر **ج** قسمت کردیم بیرون آمد **نب** وباقی ماند								ا	کح	کج	لز	ا
مکعب **ب ه لط کو کب کط کح لب نب** بگرفتیم بود		ط	یا	ح	یط	کط	ح[73]	نز	لط	مز	ن	یط
مکعب ثامن ازان اسقاط کردیم باقی ماند									یا	کد	یج	یح
آنرا بر باقی عدد افزودیم حاصل شد								ا	لط	مز	ن	یط
آنرا بر **ج** قسمت کردیم بیرون آمد **لج** وباقی ماند									ء	مز	ن	یط
وباین عمل را قطع کردیم پس خارج قسمت با شد	ب	ه	لط	کو	کب	کط	کح	لب	نب	لج		

63 مرفوع مرة است بر T6003
64 T6003 om. ان
65 T4716 om. فوق
66 T4716 اسفاط
67 T4716 بگرفتیم حاصل شد
68 T6003 اسفاط
69 T4716 ه
70 T6003, T4716 ا
71 T6003, T4716 ط
72 T6003, T4716 ا
73 T6003, T4716 ط

First digit **6** is raised twice. We divide that by **3** raised twice. **2** degrees cubed, that is **8**.	Raised	Degrees	Minutes	Seconds	Thirds	Fourths	Fifths	Sixths	Sevenths	Eighths	Ninths	Tenths
We add that to **49** which is degrees:	16	57	7	59	8	56	29	40				
We divide that by **3**. **5** comes out and there remains:	1	57	7	59	8	56	29	40				
Then we cube **2 5**, there is:		9	2	32	5							
From that, we subtract the cube of **2** i.e., **8** and there remains:		1	2	32	5							
We add that to the remaining number above there is:	1	58	10	31	13	56	29	40				
We divide that by **3**. **39** comes out and there remains:		1	10	31	13	56	29	40				
We took the cube **2 5 39**. There results:		9	11	2	32	27	43	39				
From that, we subtract the second cube, remains:			8	30	27	27	43	39				
We add that to the remaining number, results:		1	19	1	41	24	13	19				
We divide that by **3**. **26** comes out and there remains:			1	1	41	24	13	19				
We took the cube **2 5 39 26**. There results:		9	11	8	14	33	12	23	21	4	56	
From that, we subtract the third cube, remains:				5	42	5	28	44	21	4	56	
We add that to the remaining number, results:			1	7	23	29	42	3	21	4	56	
We divide that by **3**. **22** comes out and there remains:				1	23	29	42	3	21	4	56	
We took the cube **2 5 39 26 22**. There results:		9	11	8	19	22	41	6	52	15	1	56
From that, we subtract the fourth cube, remains:					4	49	28	43	31	10	5	56
We add that to the remaining number, results:				1	28	19	10	46	52	15	1	56
We divide that by **3**. **29** comes out and there remains:					1	19	10	46	52	15	1	56
We took the cube **2 5 39 26 22 29**. There results:		9	11	8	19	29	2	42	1	39	50	52
From that, we subtract the fifth cube, remains:						6	21	35	9	24	48	56
We add that to the remaining number, results:					1	25	32	22	1	39	50	52
We divide that by **3**. **28** comes out and there remains:						1	32	22	1	39	50	52
We took the cube **2 5 39 26 22 29 28**. There results:		9	11	8	19	29	8	50	27	19	59	43
From that, we subtract the sixth cube, remains:							6	8	25	40	8	51
We add that to the remaining number, results:						1	38	30	27	19	59	43
We divide that by **3**. **32** comes out and there remains:							2	30	27	19	59	43
We took the cube **2 5 39 26 22 29 28 32**. There results:		9	11	8	19	29	8	57	28	23	37	1
From that, we subtract the seventh cube, remains:							0	7	1	3	37	18
We add that to the remaining number, results							2	37	28	23	37	1
We divide that by **3**. **52** comes out and there remains:								1	28	23	37	1
We took the cube **2 5 39 26 22 29 28 32 52**. There results:		9	11	8	19	29	8	57	39	47	50	19
From that, we subtract the eighth cube, remains:									11	24	13	18
We add that to the remaining number, results								1	39	47	50	19
We divide that by **3**. **33** comes out and there remains:									0	47	50	19
We terminate this operation, then there results the quotient:	2	5	39	26	22	29	28	32	52	33		

[74] طریق دیگر برهانی که از مصنف نورانند مرقده منقول است[75] بچند وجه آنرا بیان کرده است. اما مآل همة یکی است[76] و در هر وجهی دو نوع بیان کرده یکی انکه وتر دو درجة استخراج کرده و از ان جیب یکدرجة معلوم کرده دیگر انکه جیب یکدرجة استخراج کرده بی استعانت بوتر دو درجة. و ما یک وجه از ان وجوه را بهر دو نوع ایراد کنیم چه مقصور بآن کفایت میشود.[77] اما بجهت نوع اول همان شکل که در طریق برهانی[78] اول مرسوم شد[79] اعاده کنیم.

واز نقطه ب عمود به بر اج اخراج کنیم. وچون در دو مثلث ابه جبه هر یک از دو زاویه ه قایمه است وبشکل عروس مربع اب مساوی دو مربع اه هب است و مربع بج مساوی دو مربع جه هب و اب بج بالفرض متساویند و به مشترک پس اه هج متساوی باشند. و از قاعده دوم معلوم شود که مربع اب مساوی حاصل ضرب قطر است در به. پس چون مربع اب را بر قطر قسمت کنند به معلوم شود و چون مربع به از مربع اب اسقاط کنند مربع اه باقی ماند.

و بعد از تقدیم این مقدمة میگویم که مطلوب معرفت خط اب است که وتر دو درجة است آنرا شی فرض کردیم. پس مربع او مال باشد[80] و قطر صد و بیست جزو است[81] که[82] دو مرفوع مرة بود. پس چون مربع اب که مال است[83] بر دو مرفوع مرة قسمت کنند خارج قسمت نیم دقیقة باشد از مال اعنی سی[84] ثانیة از مال.

The other demonstrative method which is related from the author (may light shine on his grave) is explained in various versions. Nevertheless the substance is all one, and in each version two types [are] explained: one [in] which the chord of two degrees [is] extracted and from that the sine of one degree [is] known; the other [in] which the sine of one degree [is] extracted without assistance from the chord of two degrees. And we demonstrate one of those versions by each of the two types, since a summary suffices for that. Nevertheless for the sake of the first type we repeat that same figure that was described in the first demonstrative method.

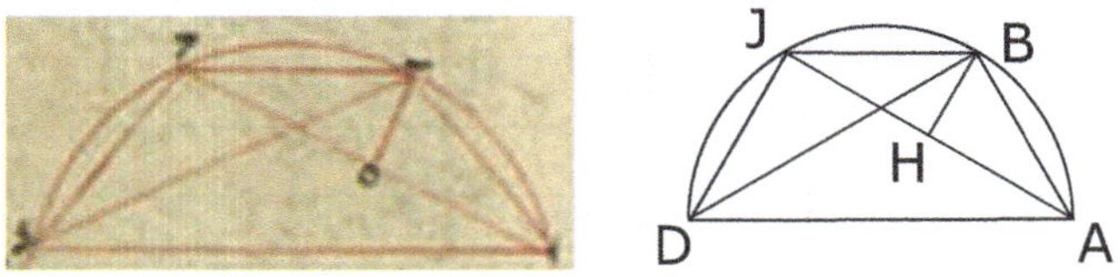

Figure 4.4: Diagram on f.33b of MS Tehran Majlis 4716 (facsimile and reconstruction)

And from point B we extend perpendicular BH upon AJ. And since in the two triangles ABH, JBH, each one of the two, angle H is right, and by the "Bride Figure" the square of AB is equal to the two squares of AH, HB; and the square of BJ is equal to the two squares of JH, HB; and AB, BJ by assumption are equal, and BH is common [to both]: therefore, AH, HJ are equal. And from the second premise it is known that the square of AB is equal to the product of multiplying the diameter by BH. So when they divide the square of AB by the diameter, BH becomes known, and when they subtract the square of BH from the square of AB, the square of AH remains.

And then from the precedence/presenting of this subject, I say that the desideratum is the acquisition of line AB, which is the chord of two degrees; we consider that the *shay*. Then its square is the *māl*, and the diameter is a hundred and twenty parts, which is two once raised. Then when they divide the square of AB, which is the *māl*, by two once raised, the quotient is half a minute of the *māl*, i.e., thirty seconds of the *māl*.

---

74 TS840 [end lacuna]
75 منقولست T6003, T4716
76 یکست T6003, T4716
77 مي شود TS840
78 برهان TS840
79 شده TS840
80 مربع مال شد T6003
81 جزوست T4716
82 که om. TS840
83 مالست T6003
84 شی TS840

و آن[85] مقدار ب ه است و[86] مربع او پانزده ثالثة باشد از مال المال. پس چون انرا[87] از مربع اب اسقاط کنند باقی ماند مربع اه یکمال[88] الا پانزده ثالثة از مال المال. و مربع اه ربع مربع اج است بشکل چهارم از ثانیة اصول. پس مربع[89] اج چهار مال باشد الا یک ثانیة از مال المال. و بشکل دوم[90] از اول مجسطی سطح اج در بد اعنی مربع اج مثل[91] مجموع سطح اب در جد و سطح ب ج در اد باشد و مقدار اد چنانچه سابقا استخراج شده و است[92] و یو مط ز نط ح نو ل سابعة.

و ب ج که مساوی اب است شی است و[93] حاصل ضرب آن در اد باشد این قدر اشیا و یو مط ز نط ح نو ل و سطح اب در جد که بهمان مربع اب است[94] یکمال[95] بود. پس چهار مال الا یک ثانیة از مال المال معادل یکمال و این قدر از شیا مذکورة باشد. و بعد از عمل جبر چهار مال معادل یکمال و یک ثانیة از مال المال و این قدر از اشیا مذکورة باشد. و بعد از عمل مقابلة سه مال معادل یک ثانیة از مال المال و این قدر از[96] اشیا مذکورة باشد. و چون[97] ثلث هر یک از متعادلی بگیرند یکمال معادل بیست ثالثة از مال المال و ثلث عدد اشیا مذکورة[98] باشد. و آن ثلث هست ب ه لو کب لط مب نح ف سابعة. و سابقا مذکور شد که اجناس متناسب اند. پس چون آنها را یک مرتبة میخط گیرند یک شی[99] معادل باشد با بیست ثالثة از مکعب و این عدد اعنی ب ه لو کب لط مب نح ف سابعة[100] پس عمل منتهی شد بشی معادل مکعب و عدد مذکور.

And that is the amount of *BH*, and its square is fifteen thirds of the *māl-māl*. Then when they subtract that from the square of *AB*, [there] remains the square of *AH*, one *māl* less fifteen thirds of the *māl-māl*. And the square of *AH* is a fourth of the square of *AJ*, by the fourth figure of the second [book] of the *Elements*. Then the square of *AJ* is four *māl* less one second of the *māl-māl*. And by the second figure of the first [book of] the Almagest, the product of *AJ* with *BD*, i.e., the square of *AJ*, is like the sum of the product of *AB* with *JD* and the product of *BJ* with *AD*, and the amount of *AD* according to the previous [discussion was] extracted, and [it] is: 6; 16, 49, 7, 59, 8, 56, 30, [to] sevenths.

And *BJ* which is equal to *AB* is the *shay*, and the product of multiplying that by *AD* is this amount of the *shay*, 6; 16, 49, 7, 59, 8, 56, 30, and the product of *AB* with *JD*, which is that very square of *AB*, is one *māl*. Then four *māl* less one second of the *māl-māl* is equal to one *māl* and this amount mentioned for the *shay*. And then from the operation of restoration, four *māl* is equal to one *māl* and one second of the *māl-māl* and this amount mentioned for the *shay*. And then from the operation of confrontation, three *māl* is equal to one second of the *māl-māl* and this amount mentioned for the *shay*. And when they take one-third of each one of the two equal [sides], one *māl* is equal to twenty thirds of the *māl-māl* and a third of the mentioned number of the *shay*. And that one-third is 2; 5, 36, 22, 39, 42, 58, 50, [to] sevenths. And [it] was previously mentioned that [variable] kinds are proportional. So when they write them [reduced] one degree, one *shay* is equal to twenty thirds of the cube and this number, i.e., 2; 5, 36, 22, 39, 42, 58, 50, [to] sevenths; then the operation is terminated with the *shay* equal to the cube and the mentioned number.

---

[85] و این T6003
[86] و om. TS840
[87] اورا TS840
[88] یک TS840
[89] مربع مال TS840
[90] دویم T6003
[91] مربع اج چهار مال باشد الا یک ثانیة از مال المال مثل T6003
[92] شده وست T6003, T4716
[93] و om. T4716
[94] مربعست T4716
[95] یک مال TS840
[96] از om. TS840
[97] چون om. T6003
[98] امذکورة TS840, T6003, T4716
[99] شی om. TS840
[100] سابعة و سابقا — سابعة om. T6003

واين از مسائل ست مشهورة[101] نيست ليكن چون عددى حاصل كنند كه تفاوت ميان او و ميان ثلث[102] عدد مذكور بيست ثالثة باشد از مكعب او آن عدد عدد مطلوب بود. پس بجهت تحصيل آن عدد[103] مكعب ثلث عدد مذكور گرفتيم بود ط ى كح ج ح نب ه لط[104] سابعة. آنرا در ك ثالثة ضرب كرديم حاصل شد ع ع ج ج كط كا ب ئز سابعة. انرا بر ثلث عدد مذكور افزوديم شد ب ه لط كو ط د ا مز مكعبشى باشد ط يا ح يو لب ل مح ط. آنرا در ك ثالثة ضرب كرديم شد[105] ع ع ج ج مب مه ل ن. آنرا بر ثلث عدد مذكور افزوديم شد[106] ب ه لط كو كب كح كط م مكعب او باشد ط يا ح يط كح نو كج ن.[107] انرا در ك ثالثة ضرب كرديم حاصل[108] شد ع ع ج ج مب مو كط لط. انرا بر ثلث عدد مذكور افزوديم حاصل شد ب ه لط كو كب كط كح كط مكعبشى بگرفتيم بود ط يا ح يط كط ط نو نا.[109] آنرا در ك ثالثة ضرب كرديم حاصل شد ع ع ج ج مب مو كط مب. آنرا بر ثلث عدد[110] مذكور افزوديم شد ب ه لط كو كب كط كح لب سابعة.

واين عدد مطلوب است[111] يعنى مقدار وتر دو درجة چه اگر مكعب آن بگيرند و در ك ثالثة ضرب كنند و بر ثلث عدد مذكور افزايند همين عدد حاصل خراج شد. پس معلوم شد كه اين عدد عدد است كه چون بيست ثالثة از مكعب او با اين[112] عدد مذكور جمع كنند مطلوب حاصل كرد.[113]

And this is not [one] of the six known problems, but when they produce a number [such] that the difference between it[self] and one-third of the mentioned number is twenty thirds of its cube, that number is the desired number. Then for the sake of producing that number, we take the cube of one-third of the mentioned number: [it] is 9; 10, 28, 3, 8, 52, 5, 39 [to] sevenths. We multiply that by 20 thirds: the result is 0; 0, 3, 3, 29, 21, 2, 57, [to] sevenths. We add that to one-third of the mentioned number: [it] is 2; 5, 39, 26, 9, 4, 1, 47; the cube-*shay* is 9; 11, 8, 16, 32, 30, 48, 9. We multiply that by 20 thirds: [it] is 0; 0, 3, 3, 42, 45, 30, 50. We add that to one-third of the mentioned number: [it] is 2; 5, 39, 26, 22, 28, 29, 40; its cube is 9; 11, 8, 19, 28, 56, 23, 50. We multiply that by 20 thirds; the result is 0; 0, 3, 3, 42, 46, 29, 39. We add that to one-third of the mentioned number; the result is 2; 5, 39, 26, 22, 29, 28, 29; we take the cube-*shay*, [it] is 9; 11, 8, 19, 29, [8], 56, 51. We multiply that by 20 thirds; the result is 0; 0, 3, 3, 42, 46, 29, 42. We add that to one-third of the mentioned number: [it] is 2; 5, 39, 26, 22, 29, 28, 32, [to] sevenths.

And this number is the desideratum, namely, the amount of the chord of two degrees, since if they take the cube of that and they multiply by 20 thirds and they add to one-third of the mentioned number, the very same number is the resulting outcome. Then it is known that this number is the number that when they add twenty thirds of its cube to this mentioned number, the desideratum results.

---

101 ازين مسايل مشهورة T6003
102 ميان او و ثلث T6003
103 تحصيل عدد T6003
104 ط ى له ج ح نب ه لط T6003
105 كرديم حاصل شد T6003
106 در — افزوديم شد om. TS840
107 ط يا ح يط كب نو كج ن T6003
108 حاصل om. T6003
109 ط يا ح يط كد ط ك نا T6003
110 عددى T6003
111 مطلوبست TS840, T6003, T4716
112 باين T6003
113 كردد TS840, T4716

و بجهت معرفت نوع دوم[114] فرض کنیم که قوسی اب جد شش درجة است بر مرکز[115] ط و هر یک از اب بج جد دو درجة است و اوتار اب اج اد بج جد وصل کنیم. پس سه وتر اول را بر نقطهای ہ ز ع تنصیف کنیم و خطوط طہ طز طع وصل کنیم و[116] هر یک عمودی[117] بود بر نظیر[118] خود بشکل ثالث از ثالثة اصول.

پس نصف قطر اط[119] وصل کنیم وآنرا بر نقطة ک تنصیف کنیم و این[120] نقطة را مرکز ساختة ببعد کا نصف دائرة رسم کنیم. و چون زوائا[121] اہط ازط اعط همة قایمة اند دایرة مذکورة بنقطهای ہ ز ع گذر و بعکس شکل سی ام[122] از ثالثة اصول و اوتار ہز زع عہ وصل کنیم و از ہ عمود ہل[123] بر وتر از[124] اخراج کنیم.

پس از بر ل نصف شود چنانکه در شکل مقدم برهن شد و چون زع منتصف[125] دو ضلع اج اد است زع موازی جد باشد بشکل دوم[126] از سادسة اصول. و این مستلزم آنست که دو مثلث ابد ازع متشابة باشند و اع نصف اد است. پس زع نصف جد بود و بمثل این بیان ہز نصف[127] بج بود و[128] اہ نصف اب است. پس اوتار اہ ہز زع همة متساوی[129] باشند.

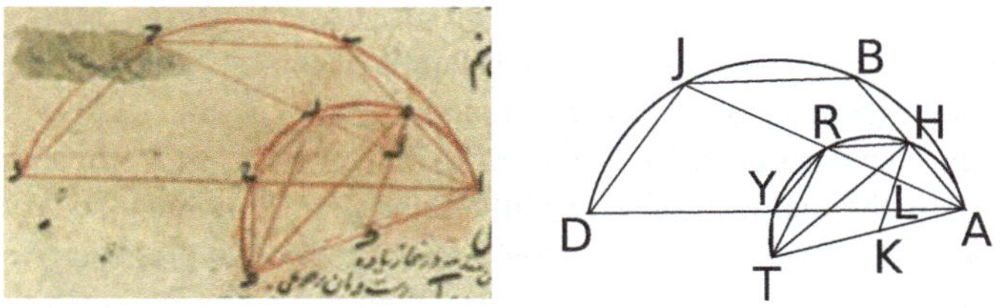

Figure 4.5: Diagram on f.34a of MS Tehran Majlis 4716 (facsimile and reconstruction)

And for the sake of obtaining the second type, we assume that an arc $ABJD$ is six degrees, on center T, and each one of [the arcs] AB, BJ, JD is two degrees, and we join the chords AB, AJ, AD, BJ, JD. Then we bisect the first three chords at the points H, R, Y, and we join lines TH, TR, TY, and each one is a perpendicular upon [its] own corresponding [chord], by the third figure of the third [book] of the *Elements*.

Then we connect half-diameter AT and we bisect that at point K, and assum[ing] this point [as] the center, with radius KA we draw a half-circle. And since angles AHT, ART, AYT all are right, the mentioned circle [is] passing through points H, R, Y, and by conversion of the thirtieth figure of the third [book of] the *Elements*; and we join chords HR, RY, YH [YH not shown in figure], and from H we extend perpendicular HL upon chord AR.

Then AR is bisected at L just as is demonstrated in the preceding figure, and since RY is the bisector of the two sides AJ, AD, RY is parallel to JD by the second figure from the sixth [book of] the *Elements*. And this is the reason that the two triangles ABD, ARY are similar and AY is half of AD. Then RY is half of JD, and by similarity of this [it is] clear [that] HR is half of BJ and AH is half of AB. Then chords AH, HR, RY all are equal.

---

114 دویم T6003
115 و بر مرکز TS840
116 و om. T6003
117 عمود TS840, T6003
118 بر قطر T6003
119 اطہ TS840
120 و آن TS840, T6003
121 زوایای T6003
122 سیم T6003
123 اہل T6003
124 از T6003
125 تنصیف T6003
126 دویم T6003
127 بیان در نصف T6003
128 و om. T6003
129 اہ ر زع جنسة متساوی T6003, اہ ہد زح همة متساوی TS840

و[130] از نقطة ک عمود ک ل بر وتر ا ر اخراج کنیم تا انرا تنصیف کند[131] بر نقطة ل و بعد از اخراج بنقطة ة رسد بقوة ثالثة اصول. پس ا ه را که جیب یکدرجة است شی فرض کنیم و ا ع که جیب سه درجة است بود ج ح کد لج نط لد کح یه سابعة. وچون مربع ا ه که مال است بر قطر دایرة صغیرة اعنی ا ط که شصت است قسمت کنند خارج قسمت یکدقیقة از مال بود. و آن مقدار ه ل است چنانکه از قاعدة دوم[132] مستفاد میشود.[133] و مربع او که[134] یک ثانیة از مال المال بود چون از مربع ا ه اسقاط کنند باقی ماند مربع ا ل یک مال[135] الا یک ثانیة از مال المال و مربع ا ل ربع مربع ا ر است.

پس مربع ا ر[136] چهار مال بود الا چهار ثانیة از مال المال و بحکم قاعدة اول سطح ا ر[137] در ه ع اعنی مربع ا ر مساوی سطح ا ه در ع ر اعنی مربع ا ه و سطح ه ر در ا ع است و مربع ا ر یک مال است[138] و سطح ه ر در ا ع این قدر از اشیا ج ح کد لج نط لد کح یه. و بعد از عمل جبر چهار مال معادل شود با یک مال[139] و چهار ثانیة از مال المال[140] و عدد اشیا مذکورة. و بعد[141] از عمل مقابلة سه مال معادل شود با چهار ثانیة از مال المال و عدد اشیا مذکورة. و[142] بعد از رو اجناس[143] متعادلة با ثلث یک مال[144] معادل شود با[145] یک ثانیة و بیست ثالثة از[146] مال المال و این قدر اشیا ا ب مح یا یط نا کط که سابعة. و بعد از خط این اجناس بیک مرتبة یک شی معادل شود با[147] یک ثانیة و بیست ثالثة از کعب و این عدد ا ب مح یا یط نا کط که.

And from point K we extend perpendicular KL upon chord AR until [it] bisects that at point L, and then from extension [it] arrives at point H, by virtue [of] the third [book of] the *Elements*. Then we assume AH, which is the sine of one degree, [to be] the *shay*, and AY, which is the sine of three degrees, is 3; 8, 24, 33, 59, 34, 28, 15, [to] sevenths. And when they divide the square of AH, which is the *māl*, by the diameter of the small circle, i.e., AT (which is sixty), the quotient is one minute of the *māl*. And that is the amount of HL, just as was understood from the second premise. And since its square, which is one second of the *māl-māl*, [is] subtract[ed] from the square of AH, [there] remains the square of AL, one *māl* less one second of the *māl-māl*; and the square of AL is one-quarter of the square of AR.

Then the square of AR is four *māl* less four seconds of the *māl-māl*, and by authority of the first premise the product of AR with HY, i.e., the square of AR, is equal to the product of AH with RY, i.e., the square of AH, and the product of HR with AY. And the square of AR is one *māl*, and the product of HR with AY is this amount of the *shay*: 3; 8, 24, 33, 59, 34, 28, 15. And then from the operation of restoration, four *māl* is equal to one *māl* and four seconds of the *māl-māl* and the mentioned number of the *shay*. And then from the operation of confrontation, three *māl* is equal to four seconds of the *māl-māl* and the mentioned number of the *shay*. And then in the manner of kinds, equalization with one-third of one *māl* is equal to one second and twenty thirds of the *māl-māl* and this amount of the *shay*: 1; 2, 48, 11, 19, 51, 29, 25, [to] sevenths. And then from the writing of these kinds [reduced] by one degree, one *shay* is equal to one second and twenty thirds from the cube and this number: 1; 2, 48, 11, 19, 51, 29, 25.

---

130 و om. T4716
131 کند T6003
132 دویم T6003
133 می شود TS840
134 و مربع انکه T6003
135 یکمال T6003
136 از T6003
137 از T6003
138 مالست T4716
139 یکمال TS840
140 از مال TS840, T4716
141 و عدد اشیا و مذکورة بعد TS840, T6003
142 و om. TS840
143 از رو ایناجناس TS840 عمل مقابلة — اجناس om. T6003
144 یکمال TS840, T6003, T4716
145 تا T6003
146 و از T6003
147 تا T6003

واین نیز منتهی شود شی معادل کعب و عدد و[148] این از مسایل ست[149] مشهورة نیست. لیکن اگر تحصیل عددی کنند که تفاوت میان آن و ثلث عدد مذکور یک ثانیة و بیست ثالثة از مکعب عدد مذکور بود آن[150] عدد مطلوب بود.

و بجهت تحصیل آن عدد مکعب ثلث عدد مذکور بگرفتیم بود ا ح مح ل کج لو ل مب یح ثامنة. آنرا در[151] ا ک[152] ثالثة ضرب کردیم و حاصل را که بود ع ع ا لا مد م لا کح ما[153] بر ثلث عدد مذکور افزودیم شد ا ب مط مج د لب ع نج ما.[154] مکعبشی باشد ا ح نج لب د ج ن نط ید نز. آنرا در ا ک ثالثة ضرب کردیم حاصل شد ع ع ا لا نا کب مه که ح.[155] انرا بر ثلث عدد مذکور افزودیم شد ا ب مط مج یا ید ید ن ح. مکعبشی باشد ا ح نج لو کو ز ع کب ی.[156] انرا در ا ک ثالثة ضرب کردیم حاصل شد ع ع ا لا نا کج ید مط ک. آنرا بر ثلث عدد مذکور افزودیم شد ا ب مط مج یا ید مد ید ک.[157] مکعبشی[158] گرفتیم بود ا ح نج لب کو ح لز ه لد.[159] انرا در ا ک ثالثة ضرب کردیم و[160] حاصلشی را[161] که بود ع ع ا لا نا کج ید نا کط بر ثلث عدد مذکور افزودیم شد ا ب مط مج یا ید مد یو کو.[162] و این عدد مطلوب است[163] چه اگر[164] مکعب این عدد را در ا ک ثالثة ضرب کنند و[165] حاصل را بر ثلث عدد مذکور افزایند همین عدد حاصل[166] آید بغیة. پس این[167] عدد جیب درجة واحد بود وچون جیب یکدرجة معلوم شد بنابر قواعد[168] سابقة جیب سایر اخرا حاصل توان کرد. و الله الموفق و المعین.[169]

And [when] even this is finished, the *shay* is equal [to] the cube and a number, and this is not among the six known problems. But if they obtain a number [for] which the difference between that and one-third of the mentioned number is one second and twenty thirds of the cube of the mentioned number, that is the desired number.

And for the sake of obtaining that number, we take the cube of one-third of the mentioned number; [it] is 1; 8, 48, 30, 23, 36, 30, 42, 18, [to] eighths. We multiply that by 1; 20 thirds, and the result, which is 0; 0, 1, 31, 44, 40, 31, 28, 41, we add to one-third of the mentioned number: [it] is 1; 2, 49, 43, 4, 32, 0, 53, 41. The cube-*shay* is 1; 8, 53, 32, 4, 3, 50, 59, 14, 57. We multiply that by 1; 20 thirds; the result is 0; 0, 1, 31, 51, 22, 45, 25, 8. We add that to one-third of the mentioned number: [it] is 1; 2, 49, 43, 11, 14, 14, 50, 8. The cube-*shay* is 1; 8, 53, 36, 26, 7, 0, 22, 10. We multiply that by 1; 20 thirds; the result is 0; 0, 1, 31, 51, 23, 14, 49, 20. We add that to one-third of the mentioned number: [it] is 1; 2, 49, 43, 11, 14, 44, 14, 20. We take the cube-*shay*: [it] is 1; 8, 53, 32, 26, 8, 37, 5, 34. We multiply that by 1; 20 thirds, and we add the result-*shay*, which is 0; 0, 1, 31, 51, 23, 14, 51, 29, to one-third of the mentioned number: [it] is 1; 2, 49, 43, 11, 14, 44, 16, 26. And this is the desired number, because if they multiply the cube of this number by 1; 20 thirds and add the result to one-third of the mentioned number, the very same number comes out, [as] desired. So this number is the sine of one degree, and when the sine of one degree is known, because of the preceding rules, [any] other sine ultimately can be produced. And it is God who prospers and assists.

---

[148] و om. T6003, T4716
[149] ست om. T6003
[150] و ان TS840
[151] ان در TS840
[152] ک T6003
[153] ع ع ا لا مد م لا لا ما T6003
[154] ا ب مط مج د لب ح نج ما T6003
[155] ع ع ا لا نا کب مه لد ح T6003
[156] ا ح نج لو لو کو ز ح لب ی T6003
[157] ا ب مط یا ید مد T6003
[158] باشد — مکعبشی om. TS840
[159] ا ح نج لب کو ح کو ح لد T6003
[160] و om. T6003
[161] را om. T6003
[162] ا ب مط مج یا مد یو کو T6003
[163] مطلوبست TS840
[164] است و اگر T6003
[165] و om. T6003
[166] همین حاصل T6003
[167] این om. T6003
[168] قاعد T6003
[169] الموفق للمرام T4716, الموفق و الموابن المعین T6003, الموفق TS840

4.3 Mathematical Notes on al-Bīrjandī's Commentary Excerpt

4.3.1 Derivation of the Cubic Equation(s) for Crd 2° or R sin 1°

According to al-Bīrjandī's remarks at the beginning of this excerpt, the geometric foundation for deriving the required cubic equation begins with constructing an isosceles trapezium as shown in figure 4.6; compare the Sanskrit version of this demonstration discussed in section 3.3.1.

The trapezium's base AD is Crd 6° and its other three sides AB, BJ, JD all have length $y = \text{Crd}\,2°$. Since any isosceles trapezium is a cyclic quadrilateral, al-Bīrjandī can invoke two relevant results about cyclic quadrilaterals. The author of the Arabic treatise states and also proves the same results more generally, deferring to a later section their application to this particular isosceles trapezium. (The Arabic text uses the term *muqaddima* for these geometrical premises; al-Bīrjandī's Persian text calls them either *muqaddima* or *qā'ida*.)

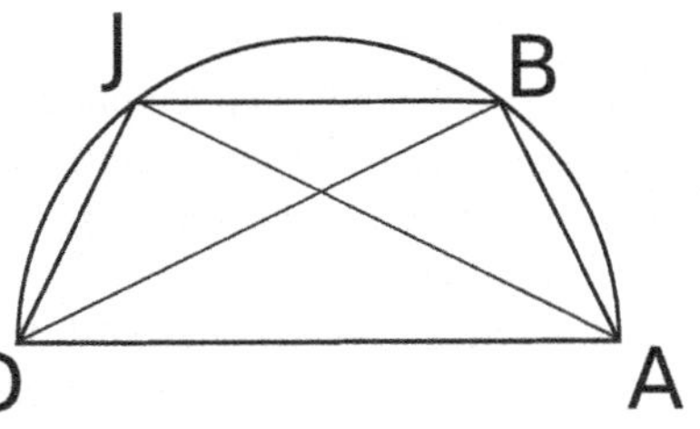

Figure 4.6: Al-Bīrjandī's trapezoidal cyclic quadrilateral on Crd 6°, redrawn from figure on f.32a of MS Tehran Majlis 4716.

First Geometrical Premise (*qā'ida awal, al-muqaddima al-awwalī*)
The first premise is the second proposition of the first book of the *Almagest*, the well-known "Ptolemy's theorem" (Toomer 1984, pp. 50–51). Namely, given a cyclic quadrilateral $ABJD$ whose diagonals are AJ and BD as shown in figure 4.6, the sum of the products of each pair of opposite sides in it is equal to the product of its diagonals:

$$AB \cdot JD + BJ \cdot AD = AJ \cdot BD. \tag{4.1}$$

In the case of our isosceles trapezium with $AD = \text{Crd}\,6°$, $AB = BJ = JD = \text{Crd}\,2° = y$, and $AJ = BD = \text{Crd}\,4°$, we can write

$$y^2 + y \cdot \text{Crd}\,6° = (\text{Crd}\,4°)^2. \tag{4.2}$$

The Arabic treatise supplies a proof of the theorem similar to Ptolemy's own; al-Bīrjandī accepts this well-known result without proof.

Second Geometrical Premise (*qā'ida dūwum, al-muqaddima al-thāniya*)
The second premise, the fourth proposition of the first book of the *Almagest*, relates the chord of the supplement of an arc α to the chord of $\alpha/2$ in a circle of radius R as follows:

$$2R - \text{Crd}(180° - \alpha) = \frac{(\text{Crd}\,\alpha/2)^2}{R}, \tag{4.3}$$

or in the case of our isosceles trapezium,

$$120 - \text{Crd}(176°) = \frac{(AB)^2}{60} = \frac{y^2}{60}, \qquad \text{or} \qquad \text{Crd}(176°) = 120 - \frac{y^2}{60}. \tag{4.4}$$

Again, a demonstration of the general theorem is supplied by the author of the Arabic exposition but not by al-Bīrjandī.

4.3.2 Algebraic Tools and Simplification of the Cubic Equation

The author of the Arabic Samarqand treatise, after proving the two general cyclic quadrilateral theorems, launches into a long passage entitled "Algebraic terms and rules". Some elements of algebra are defined, beginning with the various "kinds" (Arabic *jins*, pl. *ajnās*, a transliteration of the Greek term *genos*) or degrees of a term: the "thing" (*shay'*), the square (*māl*), the cube (*ka'b*), and the square-square (*māl al-māl*). "Additive" (*zā'id*) and "subtractive" (*nāqiṣ*) quantities and their signed arithmetic operations are also enunciated, along with the classic six types of (linear and quadratic) equations (e.g., "squares and things equal a number", and so on). The text also explains the core algebra procedures of "restoration", "confrontation", and "completion", summarized and illustrated below:

jabr restoration	$a - c = b$ $a = b + c$	*muqābala* confrontation	$a + c = b + c$ $a = b$	*takmīl* completion	$\frac{1}{c} \cdot a = b$ $a = cb$

Al-Bīrjandī states a much briefer version of this exposition preparatory to combining the two relations involving y that he derived from the two geometrical premises of section 4.3.1:

$$(\mathrm{Crd}\, 4°)^2 = y^2 + y \cdot \mathrm{Crd}\, 6° \quad \text{and} \quad \mathrm{Crd}\, 176° = 120 - \frac{y^2}{60}. \tag{4.5}$$

Invoking the Pythagorean theorem (since the inscribed triangle containing the chord of a circular arc and the chord of its supplement is right-angled with hypotenuse equal to the circle's diameter), he deduces:

$$\begin{aligned}(2R)^2 = 14400 &= (\mathrm{Crd}\, 4°)^2 + (\mathrm{Crd}\, 176°)^2 \\ 14400 &= y^2 + y \cdot \mathrm{Crd}\, 6° + \left(120 - \frac{y^2}{60}\right)^2 \\ (3,0,0)\, y &= (1,0,0) \cdot \mathrm{Crd}\, 6° + y^3.\end{aligned} \tag{4.6}$$

This cubic equation, as repeatedly emphasized by both al-Bīrjandī and the Arabic treatise, is not solvable by the ordinary methods of Islamic algebra. Hence al-Kāshī resorted instead to the approximation approach described in the next section.

4.3.3 Successive-Digits Iterative Solution of the Cubic Equation(s)

4.3.3.1 Successive-Digits Iteration for the Chord of 2°

The calculation of $y = \mathrm{Crd}\, 2°$ from the foregoing cubic equation begins by assuming knowledge of the chord of 6°, up to its sexagesimal approximation $\mathrm{Crd}\, 6° \approx 6; 16, 49, 7, 59, 8, 56, 29, 40$. Al-Bīrjandī arrives at this number by doubling the value of $R \sin 3°$ derived from the geometrically determined quantities $R \sin 18°$ and $R \sin 15°$; the Arabic treatise simply states its value. Neglecting the small imprecision in this numerical value for Crd 6°, we can rearrange the abovementioned cubic equation in y as follows:

$$\begin{aligned}(3,0,0)\, y &= y^3 + (1,0,0) \cdot \mathrm{Crd}\, 6° \\ y &= \frac{6,16,49;7,59,8,56,29,40}{3,0,0} + \frac{y^3}{3,0,0}\end{aligned} \tag{4.7}$$

The iterative solution described by both the Arabic and the Persian sources manipulates the equation's form to produce the successive sexagesimal digits of an increasingly accurate approximate value of the unknown y. In the table below, we supplement the bolded snippets of al-Bīrjandī's description of the first few steps of this procedure with more general symbolic representations (in the left column) and sample calculations (in the right). We use the same notation conventions as in our discussion of the method's Sanskrit counterpart in section 3.3.2.

"First digit 6 is raised twice. We divide that by 3 raised twice."

$$y = \frac{6,16,49;7,59,8,56,29,40}{3,0,0} + \frac{y^3}{3,0,0} \approx q_0 + r_0$$

$$y \approx q_0 + r_0 = \frac{6,0,0}{3,0,0} + \frac{16,49;7,59,8,56,29,40}{3,0,0}$$

The first approximation to the unknown y is taken to be $q_0 = 2$.

"2 degrees cubed, that is 8. We add that to 49 which is degrees."

$$\begin{aligned} y &= q_0 + r_0 + \frac{{q_0}^3}{3,0,0} + \frac{y^3 - {q_0}^3}{3,0,0} \\ &\approx q_0 + r_0 + \frac{{q_0}^3}{3,0,0} \end{aligned}$$

$$\begin{aligned} y &\approx q_0 + r_0 + \frac{{q_0}^3}{3,0,0} \\ &= 2 + \frac{16,49;7,59,8,56,29,40 + 2^3}{3,0,0} \\ &= 2 + \frac{16,57;7,59,8,56,29,40}{3,0,0}. \end{aligned}$$

"We divide that by 3. 5 comes out..."

$$q_1 + r_1 = r_0 + \frac{{q_0}^3}{3,0,0}$$

$$\begin{aligned} q_1 + r_1 &= r_0 + \frac{{q_0}^3}{3,0,0} \\ &= \frac{16,57;7,59,8,56,29,40}{3,0,0} \\ &= 0;5 + \frac{1,57;7,59,8,56,29,40}{3,0,0}. \end{aligned}$$

Second approximation $q_0;q_1 = 2;5$.

"Then we cube 2,5... From that, we subtract the cube of 2... We add that to the remaining number above..."

$$\begin{aligned} y &= q_0;q_1 + r_1 + \frac{y^3 - {q_0}^3}{3,0,0} \\ &= q_0;q_1 + r_1 + \frac{(q_0;q_1)^3 - {q_0}^3}{3,0,0} + \frac{y^3 - (q_0;q_1)^3}{3,0,0} \\ &\approx q_0;q_1 + r_1 + \frac{(q_0;q_1)^3 - {q_0}^3}{3,0,0} \end{aligned}$$

$$\begin{aligned} y &\approx q_0;q_1 + r_1 + \frac{(q_0;q_1)^3 - {q_0}^3}{3,0,0} \\ &= 2;5 + \frac{1,57;7,59,8,56,29,40 + ((2;5)^3 - 2^3)}{3,0,0} \\ &= 2;5 + \frac{1,58;10,31,13,56,29,40}{3,0,0} \end{aligned}$$

"We divide that by 3. 39 comes out..."

$$q_2 + r_2 = r_1 + \frac{(q_0;q_1)^3 - {q_0}^3}{3,0,0}$$

$$\begin{aligned} q_2 + r_2 &= r_1 + \frac{(q_0;q_1)^3 - {q_0}^3}{3,0,0} \\ &= \frac{1,58;10,31,13,56,29,40}{3,0,0} \\ &= 0;0,39 + \frac{1;10,31,13,56,29,40}{3,0,0}, \end{aligned}$$

Third approximation $q_0;q_1,q_2 = 2;5,39$.

"We took the cube 2,5,39... From that, we subtract the second cube... We add that to the remaining number..."

$$
\begin{aligned}
y &= q_0; q_1, q_2 + r_2 + \frac{y^3 - (q_0; q_1)^3}{3,0,0} \\
&= q_0; q_1, q_2 + r_2 + \frac{(q_0; q_1, q_2)^3 - (q_0; q_1)^3}{3,0,0} \\
&\quad + \frac{y^3 - (q_0; q_1, q_2)^3}{3,0,0} \\
&\approx q_0; q_1, q_2 + r_2 + \frac{(q_0; q_1, q_2)^3 - (q_0; q_1)^3}{3,0,0}
\end{aligned}
$$

$$
\begin{aligned}
y &\approx q_0; q_1, q_2 + r_2 + \frac{(q_0; q_1, q_2)^3 - (q_0; q_1)^3}{3,0,0} \\
&= 2; 5, 39 \\
&\quad + \frac{1; 10, 31, 13, 56, 29, 40 + ((2; 5, 39)^3 - (2; 5)^3)}{3,0,0} \\
&= 2; 5, 39 + \frac{1; 19, 1, 41, 24, 13, 19}{3,0,0}
\end{aligned}
$$

"We divide that by 3. 26 comes out..."

$$
q_3 + r_3 = r_2 + \frac{(q_0; q_1, q_2)^3 - (q_0; q_1)^3}{3,0,0}
$$

$$
\begin{aligned}
q_3 + r_3 &= r_2 + \frac{(q_0; q_1, q_2)^3 - (q_0; q_1)^3}{3,0,0} \\
&= \frac{1; 19, 1, 41, 24, 13, 19}{3,0,0} \\
&= 0; 0, 0, 26 + \frac{0; 1, 1, 41, 24, 13, 19}{3,0,0}
\end{aligned}
$$

Fourth approximation $q_0; q_1, q_2, q_3 = 2; 5, 39, 26$.

4.3.3.2 Successive-Digits Iteration for the Sine of 1° (in the Arabic Treatise)

A similar approximation technique for directly computing the R sin of one degree is attributed to al-Kāshī in the Arabic treatise. Its author references the same diagram that was used for the first method (Rosenfeld and Hogendijk 2003, p. 44), but the trapezium base AD is now considered equal to half Crd 6°, i.e., $R \sin 3°$ (see figure 4.7). The trapezium side AB is now half Crd 2°, i.e., the unknown $x = R \sin 1°$. Each of the diagonals AJ and BD is likewise half of Crd 4°, or $R \sin 2°$.

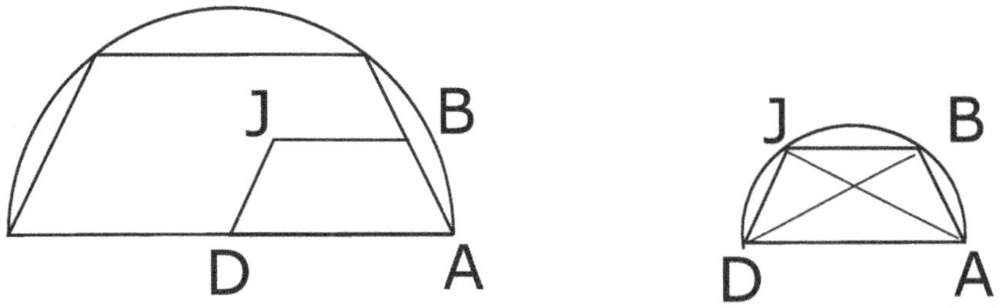

Figure 4.7: Our depiction of the half-size trapezoidal cyclic quadrilateral $ABJD$ on $R \sin 3°$. Left: Scaled from original $ABJD$ on Crd 6°, with scale factor 1/2. Right: In its own circumcircle with radius 30 units.

Al-Kāshī is alleged by the Arabic author to have reasoned as follows: By Ptolemy's theorem,

$$
x^2 + x \cdot R \sin 3° = AJ \cdot BD = (AJ)^2 = (R \sin 2°)^2. \tag{4.8}
$$

Then, invoking the aforesaid *Almagest* 1.4 relation

$$
\mathrm{Crd}(180° - \alpha) = 2R - \frac{(\mathrm{Crd}\, \alpha/2)^2}{R}, \tag{4.9}
$$

he claims[170]

$$\mathrm{Crd}(180^\circ - \widehat{AJ}) = 60 - \frac{(AB)^2}{30}, \tag{4.10}$$

correctly assuming the smaller circumradius to be 30 rather than 60.

The cubic formula involving $x = R \sin 1^\circ$ is more easily extracted from the previous one involving $y = \mathrm{Crd}\, 2^\circ = 2x$ by a simple algebra substitution, as follows:

$$\begin{aligned}(3,0,0)\, y &= y^3 + (1,0,0) \cdot \mathrm{Crd}\, 6^\circ \\ (3,0,0)\,(2x) &= (2x)^3 + (1,0,0) \cdot 2R \sin 3^\circ \\ (45,0)\, x &= x^3 + (15,0) \cdot R \sin 3^\circ \end{aligned} \tag{4.11}$$

Again neglecting the small imprecision in the numerical value of our known quantity $R \sin 3^\circ$, we can express this relation as

$$x = \frac{47,6;8,29,53,37,3,37,15}{45,0} + \frac{x^3}{45,0} \tag{4.12}$$

Laying out the first few steps of this x-iteration in the following table as we did previously for al-Bīrjandī's y-iteration, we note that the author, unlike al-Bīrjandī, rounds the remainders to a digit or so (Rosenfeld and Hogendijk 2003, pp. 46–48):

"He divided 47, raised one time, by 45, also raised one time. He obtained 1 degree and the remainder from the dividend is 2 and so on." $x = q_0 + r_0 + \frac{x^3}{45,0} \quad \approx q_0 + r_0$	$x \approx q_0 + r_0$ $= \frac{47,0}{45,0} + \frac{6;8,29,53,37,3,37,15}{45,0}$ $\approx 1 + \frac{2,0}{45,0} + \frac{6}{45,0}$ The first approximation to the unknown x is taken to be $q_0 = 1$.
"So he wrote the cube of 1 which is also 1 degree... and he added it to the remainder..." $x = q_0 + r_0 + \frac{q_0^3}{45,0} + \frac{x^3 - q_0^3}{45,0} \quad \approx q_0 + r_0 + \frac{q_0^3}{45,0}$	$x \approx q_0 + r_0 + \frac{q_0^3}{45,0}$ $= 1 + \frac{2,6 + 1^3}{45,0}$ $= 1 + \frac{2,7}{45,0}$
"Then he divided the sum, so he obtained 2 minutes, and the remainder 37;8." $q_1 + r_1 = r_0 + \frac{q_0^3}{45,0}$	$q_1 + r_1 = r_0 + \frac{q_0^3}{45,0}$ $= \frac{2,7}{45,0}$ $\approx 0;2 + \frac{37;8}{45,0}$ Second approximation $q_0; q_1 = 1;2$.

[170] The reference to the chord of the complement of "AB" in the translation in (Rosenfeld and Hogendijk 2003, p. 44) may be a typographical error for "AC" at the top of p. 59 of the facsimile edition; likewise "AB" for "AD" in note 53.

4.3.4 Ulugh Beg's Successive-Values Iteration Techniques

After describing the successive-digits iterative method for $y = \text{Crd}\,2°$ that he attributed to al-Kāshī, al-Bīrjandī goes on to explain a successive-values approach ascribed to Ulugh Beg. He notes that although there are various versions (*wajh*) of this method (*ṭarīq*), each version can be divided into two types (*naw'*): one for determining Crd 2° and the other for $R \sin 1°$. In the remainder of this chapter we summarize his demonstrations for the two types, which closely parallel the subsequent Sanskrit expositions we have analyzed in sections 3.5.1 and 3.6.1 respectively. His accompanying computations are also very similar to those discussed in sections 3.5.2 and 3.6.2, so we do not reconstruct them here.

4.3.4.1 Successive-Values Iteration for the Chord of 2°

Al-Bīrjandī commences his explanation of the first type by referencing the diagram shown in figure 4.8. This represents the original isosceles trapezium of figure 4.6 with base Crd 6° and side Crd 2°, but now with perpendicular BH dropped on the diagonal AJ. Since the two right triangles ABH and JBH have equal hypotenuses and a common side BH, their other sides AH and HJ are also equal, which al-Bīrjandī infers from the "Bride Figure" (that is, the "bride's chair" or Pythagorean theorem). He then invokes the second premise in support of the claim that

$$(AB)^2 = (2R) \cdot BH, \quad \text{and consequently} \quad BH = \frac{(AB)^2}{2R} = \frac{(AB)^2}{2,0}. \tag{4.13}$$

(More precisely, this deduction depends on an intermediate step in the well-known demonstration of the second premise whose Sanskrit version we discussed in section 3.3.1. Namely, because right triangle ABH is similar to another right triangle whose hypotenuse is the circle diameter and whose shortest side is AB, it must follow that $\frac{AB}{2R} = \frac{BH}{AB}$.)

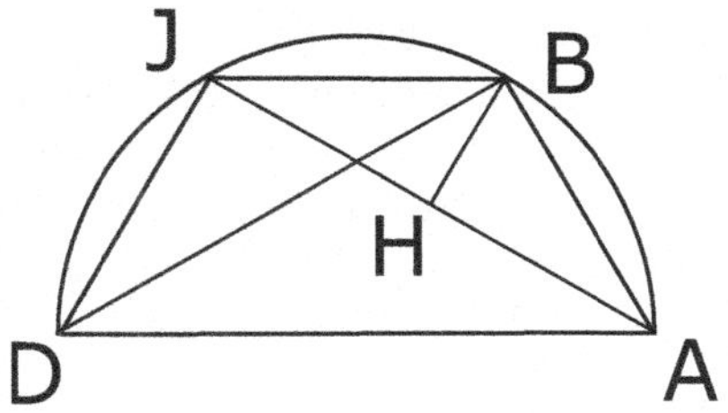

Figure 4.8: Redrawing of al-Bīrjandī's figure on f.33b of MS Tehran Majlis 4716.

Recollecting that AB is the unknown y, by the Pythagorean theorem

$$(AH)^2 = (AB)^2 - (BH)^2 = (AB)^2 - \left(\frac{(AB)^2}{2,0}\right)^2 = y^2 - \left(\frac{y^2}{2,0}\right)^2 = y^2 - (0;0,0,15)\,y^4. \tag{4.14}$$

Since AH is half of the chord AJ, the square of the former is one-fourth the square of the latter (which al-Bīrjandī justifies by appeal to *Elements* 2.4). Consequently,

$$(AJ)^2 = 4(AH)^2 = 4y^2 - (0;0,1)\,y^4. \tag{4.15}$$

And by Ptolemy's theorem (*Almagest* 1.2) as before,

$$(AJ)^2 = (AB)^2 + AB \cdot \text{Crd}\,6° = y^2 + y \cdot \text{Crd}\,6°, \tag{4.16}$$

so

$$
\begin{aligned}
4y^2 - (0;0,1)\,y^4 &= y^2 + y\cdot \mathrm{Crd}\,6^\circ \\
4y^2 &= y^2 + (0;0,1)\,y^4 + y\cdot \mathrm{Crd}\,6^\circ && \text{(restoration)}\\
3y^2 &= (0;0,1)\,y^4 + y\cdot \mathrm{Crd}\,6^\circ && \text{(confrontation)}\\
y^2 &= (0;0,0,20)\,y^4 + y\cdot \frac{\mathrm{Crd}\,6^\circ}{3}\\
y &= (0;0,0,20)\,y^3 + \frac{\mathrm{Crd}\,6^\circ}{3} && = (0;0,0,20)\,y^3 + 2;5,36,22,39,42,58,50,
\end{aligned}
\tag{4.17}
$$

neglecting the inaccuracy of the constant term.

Al-Bīrjandī notes that a numerical value that satisfies the equation to the calculator's desired precision is an acceptable solution for y, even though the analytic solution of the cubic is not available. Such a value can be obtained by means of a fixed-point algorithm equivalent to the following iterative equation:

$$
y_{k+1} = y_0 + (0;0,0,20)\,(y_k)^3 \tag{4.18}
$$

with y_0 taken to be one-third of the "mentioned number" or Crd 6°. From this starting-point, al-Bīrjandī carries out the iteration for four cycles before arriving at a fixed value of y.

4.3.4.2 Successive-Values Iteration for the Sine of 1°

Al-Bīrjandī's description of the "second type" of this technique, for approximating $R\sin 1^\circ$, begins with a more elaborate version of the same isosceles-trapezium diagram (see figure 4.9). In the diagram, assuming that the center of the circumcircle of the trapezium $ABJD$ is T, we draw radii from T perpendicularly bisecting chord AD at Y, chord AJ at R, and chord AB at H. Connecting another radius AT, we make three right triangles AHT, ART, AYT all lying on the same hypotenuse AT, which we bisect at K. The semicircle with diameter AT thus must pass through the rectangular-vertex points at H, R and Y. Al-Bīrjandī notes, for reasons not entirely clear to us, that this result is justified by "conversion" (*ʿaks*) of *Elements* 3.30, which shows that the perpendicular bisector of a semicircle's diameter also bisects its circumference. (He may have intended to refer instead to the converse of the proposition designated 3.31 in modern editions of the *Elements*, a version of Thales' theorem stating that a triangle inscribed in a semicircle is right-angled.)

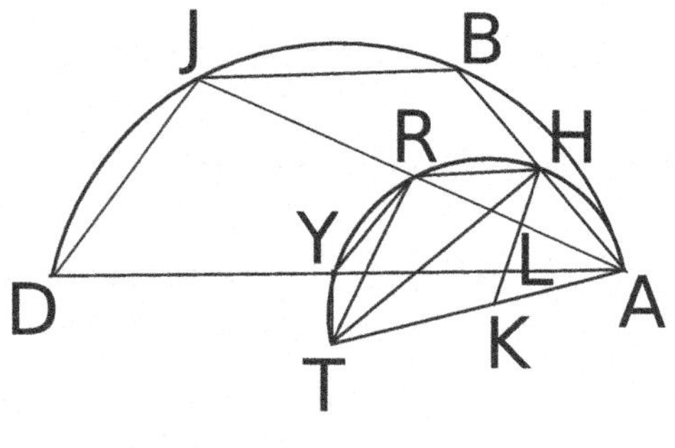

Figure 4.9: Redrawing of al-Bīrjandī's figure on f.34a of MS Tehran Majlis 4716.

Joining chords HR and RY in this smaller semicircle, we produce triangles AHR and ARY, which are respectively similar to triangles ABJ and AJD, by *Elements* VI.2 on proportional cutting of triangle sides. Al-Bīrjandī also prescribes here joining chord YH (apparently somewhat unnecessarily) and bisecting AR at L. He likewise notes that radius KLH is the perpendicular bisector of AR, by (an unspecified proposition in) "the third book of the *Elements*".

The key feature of this construction is the isosceles trapezium $AHRY$ similar to $ABJD$, with scale factor 1/2. Its circumcircle radius AK is likewise half the radius AT of the circumcircle of $ABJD$. So we can denote the lengths of its sides and diagonals, whose counterparts in $ABJD$ were defined to be the Chords of various arcs, as *R*sines of the corresponding half-arcs. Thus we may write

$$
AH = \frac{1}{2}AB = \frac{1}{2}\mathrm{Crd}\,2^\circ = R\sin 1^\circ = x \quad \text{and} \quad AY = \frac{1}{2}AD = \frac{1}{2}\mathrm{Crd}\,6^\circ = R\sin 3^\circ = 3;8,24,33,59,34,28,15,
$$

up to the accuracy of this numerical value.

The same relations that we deduced geometrically among segments AB, AJ and BH in figure 4.8 likewise hold among the corresponding segments AH, AR and HL in figure 4.9. Arguing from the same geometric premises as before, we can turn them into symbolic equations, this time in the unknown $AH = x$ with radius $R = \frac{1}{2}(1,0)$:

$$\begin{aligned} HL = \frac{(AH)^2}{2R} = \frac{x^2}{1,0} &= (0;1)\,x^2 \\ (AH)^2 - (HL)^2 = (AL)^2 &= \left(\frac{AR}{2}\right)^2 \\ (AH)^2 - (HL)^2 &= x^2 - (0;0,1)\,x^4 \\ (AR)^2 &= 4x^2 - (0;0,4)\,x^4 \end{aligned} \tag{4.19}$$

and by Ptolemy's theorem,

$$(AR)^2 = (AH)^2 + AH \cdot R\sin 3^\circ = x^2 + x \cdot R\sin 3^\circ, \tag{4.20}$$

so

$$\begin{aligned} 4x^2 - (0;0,4)\,x^4 &= x^2 + x \cdot R\sin 3^\circ \\ 4x^2 &= x^2 + (0;0,4)\,x^4 + x \cdot R\sin 3^\circ && \text{(restoration)} \\ 3x^2 &= (0;0,4)\,x^4 + x \cdot R\sin 3^\circ && \text{(confrontation)} \\ x^2 &= (0;0,1,20)\,x^4 + x \cdot \frac{R\sin 3^\circ}{3} \\ x &= (0;0,1,20)\,x^3 + \frac{R\sin 3^\circ}{3} \\ &= (0;0,1,20)\,x^3 + 1;2,48,11,19,51,29,25, \end{aligned} \tag{4.21}$$

again neglecting the inaccuracy of the constant term.

Al-Bīrjandī therefore prescribes a fixed-point algorithm equivalent to the following iterative equation:

$$x_{k+1} = x_0 + (0;0,1,20)\,x_k{}^3 \tag{4.22}$$

with x_0 equal to one-third of the "mentioned number", in this case $R\sin 1^\circ$. Four iterations produce a fixed value for x, which concludes al-Bīrjandī's discussion of these methods.

Chapter 5
Iterative Sine 1° Computations in the Sanskrit Tradition

5.1 Nityānanda's Iterative Methods for the Sine of 1°

As previously noted in section 1.2, the iterative-approximation approach discovered by al-Kāshī for computing the sine of one degree appears to have entered the Sanskrit mathematical corpus via the 1639 *Sarvasiddhāntarāja* of Nityānanda. This comprehensive treatise devotes a section of its trigonometry exposition to $R \sin 1°$ methods (verses I.3.60–85) (Montelle and Ramasubramanian 2018). Here we describe the parts of that section that are closely related to the earlier Arabic and Persian material summarized by al-Bīrjandī, and the later Sanskrit work of Jagannātha and Nayanasukha at the Jaipur court.

5.1.1 Approximation Algorithms Used by Nityānanda

5.1.1.1 Successive-Digits Iteration for Twice $R \sin 1°$

The first procedure described in this *Sarvasiddhāntarāja* section (vv. 60–63ab) iteratively solves a cubic equation for twice $R \sin 1°$ (here denoted y as usual), though Nityānanda notes that the technique applies more generally to finding $2\,R \sin(\alpha/3)$ for any angle α whose $R \sin$ is known:

$$y = \frac{2R \sin 3°}{3} + \frac{y^3}{3R^2}. \tag{5.1}$$

This is fundamentally the same as the cubic equation in y discussed in sections 3.3 and 4.3.3. But in this variant a generalized trigonometric radius R now replaces the standard Greco-Islamic $R = 1,0 = 60$, as it does in the computation of $Y = 2\,R \sin(\alpha/3)$ described in section 3.1. The solution of this cubic likewise is equivalent to the corresponding successive-digits iteration for Crd 2° discussed in those previous sections.

5.1.1.2 Successive-Values Iterations for Twice $R \sin 1°$ and for $R \sin 1°$

Nityānanda's next algorithm (vv. 63cd–64ab) generates successive approximations to the whole value of y in the above cubic equation, rather than its successive sexagesimal digits. Taking $y_0 = \frac{2}{3} R \sin 3°$, the algorithm produces

$$y_{k+1} = y_0 + \frac{(y_k)^3}{3R^2}, \tag{5.2}$$

corresponding to the y-version of the familiar iterative equation 1.5 with generalized R-value.

C. Montelle et al., *From Samarqand to Jaipur*, Sources and Studies in the History of Mathematics and Physical Sciences, https://doi.org/10.1007/978-3-031-89087-1_5

In the following technique (vv. 64cd–65ab) Nityānanda prescribes[1] a procedure mirroring the x-version of the same iterative equation, again generalized to any value of R. Namely, taking $x_0 = \frac{1}{3}R\sin 3°$, one computes

$$x_{k+1} = x_0 + \frac{4(x_k)^3}{3R^2}. \tag{5.3}$$

5.1.2 Demonstrations and Auxiliary Material

The remaining verses in this section of the *Sarvasiddhāntarāja* treat the following topics:

- The numerical value 3; 8, 24, 33, 59, 34, 28, 14, 50 for $R\sin\ 3°$.
- A Euclidean-style geometrical construction of a letter-labeled diagram representing the cyclic quadrilateral, an isosceles trapezium with three equal sides, on which the relation between the unknown chord y and its cube depends, and its diagonals. The diagram contains, besides the fundamental trapezium drawn in figure 4.8, some of the additional lines and circumcircle center shown in figure 4.9.
- Algebraic derivations for two expressions for the square of the diagonal of this cyclic quadrilateral, and consequently for the cubic equation relating the unknown chord to its cube.
- The working out of the above derivations in algebra notation.
- Trigonometric determination of the $R\sin$ of one arcminute from the $R\sin$ of one degree.

5.2 Iterative Sine Computations Described by Jagannātha

As discussed in section 1.2, Nityānanda's *Sarvasiddhāntarāja* was one of the sources employed in the composition of Jagannātha Samrāṭ's *Siddhāntakaustubha* (*SK*), of which the initial verse-only version was apparently composed around 1727. Either Jagannātha or his colleague Nayanasukha, or both, revised and expanded it with the addition of prose portions in the 1730s. This expanded version in at least two different recensions was incorporated into Jagannātha's *Siddhāntasamrāṭ* (J_S). Most of the content in its *jyācāpa* section relating to iterative sine computations is included in the excerpt in the Jaipur manuscript already examined in chapters 2 and 3. In the present section we survey the additional material in the *Siddhānta-kaustubha*/*Siddhāntasamrāṭ* addressing the problem of iterative sine computation.

The core of this material is Jagannātha's *jyācāpa* verses 14–23 in the *Siddhāntasamrāṭ* recension (Sharma 1967–1969, pp. 1050–1051), corresponding to verses II.1.62–71 in the earlier *Siddhāntakaustubha* recension (Pingree 2004, S46–S47). These verses expand and adapt Nityānanda's formulas for computing $R\sin 1°$, analyzed in section 5.1. Jagannātha's versions generate the sine or double sine of an arbitrary arc $\alpha/3$ given that $R\sin\alpha$ is known. The similarity between Nityānanda's and Jagannātha's verses in the ordering, and occasionally in the wording, of their rules reinforces the hypothesis that the versified *Siddhāntakaustubha* material was adapted directly from its *Sarvasiddhāntarāja* counterpart.

Section 5.2.1 transcribes and translates the verses *SK* II.1.48–71/J_S 1–23. The mathematical content of the core verses relating to $R\sin 1°$ is summarized in section 5.2.2. (Trivial differences in sandhi between the two published versions, including use of *anusvāra* and *avagraha*, are not noted in the transcription.) In section 5.2.3, we transcribe from the *Siddhāntasamrāṭ* (Sharma 1967–1969, pp. 1074–1076) and translate the additional prose material in the *jyācāpa* section following the end of the excerpt in the Jaipur manuscript. (The transcription of the prose section follows the irregular sandhi of the published text, except where we have noted proposed corrections for the sake of the sense.) Its mathematical content is discussed in sections 5.2.4 and 5.2.5.

---

[1] The latter part of the translation in (Montelle and Ramasubramanian 2018, p. 14) of the following lines of Nityānanda is slightly misleading:

tribhāgajīvātrilavaḥ pṛthaksthaḥ *svatryaṃśayuktena ghanena tasya* || *tribhajyakāvargahṛtena yuktaḥ*	A third part of the Sine of three degrees [is obtained and] is put down separately. Then, one third of the cube of this is divided by the square of the radius and this is added to [the value which was put down separately, i.e., Sine of three degrees divided by 3].

A more accurate rendering (using our standard notation $R\sin$ in place of the equivalent capitalized Sine) would be:

> A third part of the $R\sin$ of three degrees, put down separately, [is] increased by the cube of that [quantity] increased by its own third part [and] divided by the square of the radius.

5.2.1 Jagannātha's *jyācāpa* Verses *SK* II.1.48–71/J_S 1–23

अथ ज्योत्पत्तिः ॥

Now the sine-determination.

ज्योत्पत्तिं ये विज्ञानन्त त एवाचार्यसंज्ञिताः ॥
अतस्तां वच्मि विविधां गणकानन्दकारिणीम् ॥ ४८/१ ॥

Only those who understand the sine-determination are called "professors" (*ācārya*). So I will state [its] various forms [that] delight mathematicians. (48/1)

व्यासार्धमानेन विधाय वृत्तं
समावनौ चक्रकलाङ्कितं च ॥
दिगङ्कितं प्रागपराख्यसूत्र-
युक्तं भुजांशानुभयत्र दत्त्वा ॥ ४९/२ ॥

Making on level ground a circle with the measure of the radius, marked with the arcminutes of the circle, marked with the [cardinal] directions [and] combined with a [diameter] line called "east-west", and setting on both sides [of the east-west line] the degrees of the arc: (49/2)

तदग्रयोर्यद्धि निबद्धसूत्रं
पूर्णज्यका द्विघ्नभुजांशकानाम् ॥
ज्यां ज्यार्धमेव[2] प्रवदन्ति तज्ज्ञा
यतस्तदर्धाग्रगतो ग्रहेन्द्रः ॥ ५०/३ ॥

the line connecting their two endpoints is the full chord (*pūrṇa-jyā*) of the degrees of double the arc. [The ones] knowledgeable about it assert that "*jyā*" [means] just the half-chord (*jyā-ardha*), since [in spherical astronomy] planets etc. [are considered placed] at the tip of that half[-chord]. (50/3)

अतो वृतेः प्राक्परचिह्नतश्च
भुजांशकानेकत एव दत्त्वा ॥
तिर्यग्गता[3] मध्यमसूत्रतश्च
रेखा भवेद्या भुजसिञ्जिनोह ॥ ५१/४ ॥

So, setting the degrees of the arc on just one side with respect to the chosen east [or] west point, the line that is perpendicular to the central [diameter] line will be the sine of the arc. (51/4)

एवं हि याम्योत्तरसूत्रतश्च
कोट्यंशसक्ता खलु कोटिजीवा ॥
तद्वर्गयोगस्य पदं त्रिभज्या
कर्णो भवेत्क्षेत्रमिदं[4] हि जात्यम् ॥ ५२/५ ॥

And in the same way [the line] attached at the degrees of the complement-arc [and perpendicular] to the south-north line is the complement-sine (*koṭi-jīvā*). The square-root of the sum of their squares will be the radius [*tribhajyā*, lit. "sine of three [zodiacal] signs"], the hypotenuse: this figure is a right triangle (*jātya*). (52/5)

त्रिज्याकृतिर्दोर्गुणवर्गहीना
तस्याः पदं कोटिलवज्यका स्यात् ॥
कोटिज्यकावर्गविहीनिता या-
स्त्रिज्याकृतेः स्याच्च पदं भुजज्या ॥ ५३/६ ॥

The square of the radius is decreased by the square of the sine: its square-root is the cosine. And the square-root of the square of the radius decreased by the square of the cosine is the sine of the arc. (53/6)

दोर्ज्या विशुद्धा त्रिभसिञ्जिनीतः
कोट्युत्करमज्या विशिखाभिधाना ॥
कोटिज्यकोना यदि भत्रयज्या
भुजोत्क्रमज्या प्रभवेत्तदानीम्[5] ॥ ५४/७ ॥

The sine subtracted from the radius is the versine of the complement, called the "arrow"; when the radius is decreased by the cosine, then [there] results the versine of the arc. (54/7)

क्रमोत्क्रमज्ये[6] भुजकोटिरूपे
तद्वर्गयोगाब्धिलवस्य मूलम् ॥
भवेत्तदर्धांशकसिञ्जिनी[7] वा ॥ ५५/८a–c ॥

The regular and versed sines [have respectively] the form of the sine and cosine. The square-root of the fourth part of the sum of their squares is just the sine of the degrees of half of that [arc]. (55/8 a–c)

---

2 ज्यार्द्धनेव (Pingree 2004, S44)
3 तिर्यग्मता (Sharma 1967–1969, p. 1049)
4 भवेल्क्षेत्र० (Pingree 2004, S44)
5 प्रमेत्तदानीं (Sharma 1967–1969, p. 1049)
6 क्रमोत्क्रमे ज्ये (Sharma 1967–1969, p. 1049)
7 भवेद् भुजार्द्धांशकसिंजिनी (Sharma 1967–1969, p. 1049)

त्रिज्योत्क्रमज्यार्धहतेस्तु मूलम् ॥ ५५d SK ॥

[Only in *SK*:] [So is] the square-root of the product of the radius [with] half the versine. (55d)

प्रकारतस्तां प्रवदामि सम्यक् ॥
व्यासोत्क्रमज्याहतितूर्यभाग-
मूलं तदर्धांशकसिञ्जिनी वा ॥ ८d J_S ॥

[Only in J_S:] I will state that same [result] in [another] way. The square-root of a fourth part of the product of the diameter and the versine is just the sine of the degrees of half of that [arc]. (8d)

व्यासाच्छरघ्नाच्छरवर्गहीना-
न्मूलं भुजज्याथ तदीयवर्गे ॥
शरोद्धृताढ्ये खलु विस्तृतिः स्यात्
तदर्धकोट्यन्तरतः शरः स्यात् ॥ ५६/९ ॥

The square-root from the radius multiplied by the versine [and] decreased by the square of the versine is the sine of the arc. When its square is divided [and] increased by the versine, [it] is indeed the diameter. The versine is the difference of half of that and the cosine. (56/9)

व्यासार्धमानं त्रिभसिञ्जिनी स्याद्[8]
भवेत्तदर्धं खगुणांशजीवा ॥ ५७/१० ab ॥

The amount of the radius is the sine of three [zodiacal] signs [i.e., 90°]; half of that is the sine of thirty degrees. (57/10 ab)

तत्कोटिजीवा खरसांशकज्या
त्रिज्यार्धमूलं शरवेदजीवा ॥ ५७cd SK ॥

[Only in *SK*:] The cosine of that is the sine of sixty degrees; the square-root of half [the square of] the radius is the sine of forty-five. (57 cd)

त्रिज्याकृतेरर्धपदं तु यत्स्यात्
तद्बाणवेदप्रमितांशजीवा ॥ १०cd J_S ॥

[Only in J_S:] Whatever is the square-root of half of the square of the radius is the sine of degrees equal to forty-five. (10 cd)

त्रिज्यार्धवर्गाच्छरसङ्गुणाच्च[9]
मूलं खरामांशकजीवयोनम् ॥
तदर्धकं स्याद्धृतिभागजीवा[10]
तत्कोटिजीवा द्विनगांशकानाम् ॥ ५८/११ ॥

And the square-root from the square of half the radius multiplied by five is decreased by the sine of thirty degrees. Half of that is the sine of eighteen degrees. Its cosine is [the sine] of seventy-two degrees. (11/58)

तस्यास्तदर्धांशकसिञ्जिनी स्यात्
तत्कोटिजीवा कृतमार्गणज्या ॥
पुनस्ततस्तद्दलभागकानां
जीवा प्रसाध्या नगदस्रकाणाम् ॥ ५९/१२ ॥

From that is [derived] the sine of half its degrees [i.e., 36°]; the cosine of that is the sine of fifty-four. Again, the sine of half its degrees, twenty-seven, is obtained from that. (59/12)

अथ द्विगुणांशजीवा ॥

Now the sine of twice the degrees.

दोःकोटिजीवाभिहतिर्विभक्ता
राशिज्यया स्याद्विगुणांशजीवा ॥
तत्कोटिजीवा भुजकोटिजीवा-
वर्गान्तरं त्रिज्यकया विभक्तम् ॥ ६०[11] ॥

[Only in *SK*:] The product of the sine and cosine, divided by the sine of a [zodiacal] sign [i.e., 30°], is the sine of twice the degrees. The cosine of that is the difference of the squares of the sine and cosine, divided by the radius. (60)

अन्योन्यकोटिज्यकया विनिघ्नौ[12]
भुजांशजीवे त्रिभजीवयाप्ते ॥
तल्लब्धयोगान्तरके क्रमेण
चापैक्यचापान्तरजीवके स्तः ॥ ६१/१३ ॥

Two [distinct] sines of degrees of arc are multiplied by each other's cosine [and] divided by the radius. The sum and difference of those quotients are the sines of, respectively, the sum and difference of the arcs. (61/13)

[8] ०सिञ्जिनीह (Sharma 1967–1969, p. 1049)
[9] ०वर्गादिषुसंगुणाच्च (Sharma 1967–1969, p. 1050)
[10] स्याद्धृतिनागजीवा (Pingree 2004, S45)
[11] om. (Sharma 1967–1969, p. 1050)
[12] ०विनिघ्नो (Sharma 1967–1969, p. 1050)

अथ चापज्याज्ञाने सति चापतृतीयांशज्यानयनम् ॥

Now the derivation of the sine of a third part of the arc, with knowledge of the sine of the arc [given].

दोर्ज्यां विधाय यमसङ्गपितां[13] त्रिभक्तां
कृत्वा पृथक्फललवान् विधीत धीमान्[14] ॥
व्यासार्धवर्गविहृते घनके तु तेषां
भागावशेषसहितं फलमग्निभक्तम् ॥ ६२/१४ ॥

Having determined the sine of the arc and made [it] multiplied by two and divided by three, the wise one should place [? reading *dadhīta* for *vidhīta*] the degrees of the result separately. But when the cube is divided by the square of the half-diameter, the result increased by the remainder from [dropping] the degrees is divided by three. (62/14)

लब्धं कला लवतले विनिवेश्य पङ्क्तेः
कार्यो घनः प्रथमकेन घनेन हीनः ॥
व्यासार्धवर्गविहृतः फलयुक्तशेषं
रामैर्भजेद्विकलिकास्तदधो निवेश्याः[15] ॥ ६३/१५ ॥

[When one has] entered the quotient (in minutes) below the degrees in the *paṅkti*, the cube is to be made, diminished by the first cube [and] divided by the square of the half-diameter. One should divide by three the remainder increased by the result. The seconds [of the quotient] are set down beneath that. (63/15)

पङ्क्तेर्घनात् पूर्वघने विशोधिते
त्रिभज्यकावर्गहृते तु यत् फलम् ॥
भाज्यावशेषे सहितं विधाय
मुहुस्तदग्रे तु विधिः पुरोक्तः ॥ ६४/१६ ॥

When the cube of the previous [*paṅkti*] is subtracted from the cube of the [new] *paṅkti* and divided by the square of the radius [lit. "sine of three signs"], whatever is the result, [when one has] determined [that] added to the remainder from the division, the rule [is to be carried out] repeatedly as previously stated, but at the end of that [*paṅkti*]. (64/16)

पङ्क्त्यर्धकं स्याद्धनुषस्त्रिभाग-[16]
जीवाथ तां वच्मि पुनः प्रकारात् ॥
भेदैरनेकैर्गणकस्य बुद्धि
प्रवीणतामेति यतः स्वशास्त्रे ॥ ६५/१७ ॥

Half the [final] *paṅkti* is the sine of a third part of the arc. Now furthermore, I state that according to [another] method, since the wisdom of the mathematician attains skill in his own discipline by means of many variations. (65/17)

स्वत्र्यंशहीना भुजसिञ्जिनी वा
पृथग्घनो ऽस्यास्त्रिगुणस्य कृत्या ॥
त्रिघ्न्या हृतो लाब्धियुता मुहुर्वा
तदर्धकं चापगुणांशजीवा ॥ ६६/१८ ॥

Alternatively: The sine of the arc is diminished by its own third part. The cube of that [placed] separately is divided by three times the square of the radius. [The diminished sine] is added to the quotient, [and] thus repeatedly. Half of that is the sine of a third part of the arc. (66/18)

भुजांशजीवात्रिलवः पृथक्स्थः
स्वत्र्यंशयुक्तेन घनेन तस्य ॥
त्रिभज्यकावर्गविभाजितेन
युक्तो मुहुः स्यात् त्रिलवस्य जीवा ॥ ६७/१९ ॥

A third part of the sine of the degrees of the arc, standing separately, repeatedly increased by its cube [which is] increased by its own third part [and] divided by the square of the radius, is the sine of the third part [of the arc]. (67/19)

---

[13] दोर्ज्याविधायकरसंगुणितां (Sharma 1967–1969, p. 1050)

[14] कुर्यातपृथक् फललवान् गणकप्रवीणः (Sharma 1967–1969, p. 1050)

[15] ०कालिका० (Sharma 1967–1969, p. 1050)

[16] ०स्याद्धि धनुस्त्रि० (Sharma 1967–1969, p. 1050)

पुनः प्रकारैर्वक्ष्यामि चापत्र्यंशस्य सिञ्जिनीम्[17] ॥
खमभ्रं खं नखा ० ।० ।० ।२० श्चायं गुणो रूपघनस्य
च ॥ ६८/२० ॥

Again, I will state the sine of a third part of the arc by means of [different] methods. Zero, zero, zero and twenty [0, 0, 0, 20]: and this [is] the multiplier of the cube of the number. (68/20)

भुजज्यका स्वत्रिलवेन हीना
घनेन तस्या गुणकाहतेन ॥
युता मुहुस्तद्दलकं ज्यका स्याद्
धनुस्त्रिभागस्य पुरोक्ततुल्या ॥ ६९/२१ ॥

The sine of the arc diminished by its own third part [is] repeatedly increased by its cube multiplied by the multiplier. Half of that is the sine of a third part of the arc, equal to the previously stated [value]. (69/21)

प्रकारान्तरतो वक्ष्ये चापत्रिलवमौर्विकाम् ॥
रूपाणां गुणकश्चायं खमाकाशं धरा नखाः
० ।० ।१ ।२० ॥ ७०/२२ ॥

I shall explain by another method the sine of a third part of the arc. And this is the multiplier of the numbers: zero, zero, one, twenty [0, 0, 1, 20]. (70/22)

भुजांशमौर्वीत्रिलवः पृथक्स्थो
घनेन तस्य स्वगुणाहतेन ० ।० ।१ ।२० ॥
युतो ऽसकृद्धा धनुषस्त्रिभाग-
मौर्वी भवेत् प्राग्ज्यकया समाना ॥ ७१/२३ ॥

Alternatively: a third part of the sine of the degrees of the arc, standing separately, is increased repeatedly by its cube multiplied by its own multiplier 0, 0, 1, 20. The sine of a third part of the arc becomes equal to the previous sine. (71/23)

[17] सिञ्जिनी (Pingree 2004, S47), सिंजिनीं (Sharma 1967–1969, p. 1051)

5.2.2 Mathematical Notes on Jagannātha's *jyācāpa* Verses *SK* II.1.62–71/J$_S$ 14–23

The initial verses of this section, *SK* II.1.48–61/J$_S$ 1–13, deal with fundamental trigonometry facts and identities that are not analyzed in detail here. They were apparently inspired by, and to some extent recapitulate, verses I.3.19–59 in the trigonometry exposition of Nityānanda's *Sarvasiddhāntarāja*. For a detailed discussion of that material, see (Montelle and Ramasubramanian 2018), whose translations of Nityānanda's verses have helped illuminate Jagannātha's.

5.2.2.1 Successive-Digits Iteration for Twice $R\sin(\alpha/3)$

The solution technique described in verses *SK* II.1.62–65/J$_S$ 14–17 uses a generalized form of Nityānanda's successive-digits iteration for $Y = 2R\sin(\alpha/3)$. Its procedure was discussed above in section 3.1.

5.2.2.2 Successive-Values Iterations for Twice $R\sin(\alpha/3)$ and for $R\sin(\alpha/3)$

The four successive-values iterations described in verses *SK* II.1.66–71/J$_S$ 18–23 are likewise generalized and modified variants of Nityānanda's successive-values iterations for $R\sin\ 1°$ or its double. If $Y = 2R\sin(\alpha/3)$ and $X = Y/2 = R\sin(\alpha/3)$ are expressed by the usual cubic relations:

$$Y = \frac{2R\sin\alpha}{3} + \frac{Y^3}{3R^2}, \qquad X = \frac{R\sin\alpha}{3} + \frac{4X^3}{3R^2},$$

then Jagannātha's algorithms reduce to the following iterative equations.

Verse 66/18. Assuming $Y_0 = \frac{2}{3}R\sin\alpha$,

$$Y_{k+1} = Y_0 + \frac{(Y_k)^3}{3R^2}. \tag{5.4}$$

Verse 67/19. Assuming $X_0 = \frac{1}{3}R\sin\alpha$,

$$X_{k+1} = X_0 + \frac{(X_k)^3 \cdot 4/3}{R^2}. \tag{5.5}$$

The following two rules require the additional assumption $R = 60$, although this is not stated by Jagannātha.

Verses 68–69/20–21. Assuming $Y_0 = \frac{2}{3}R\sin\alpha$ and the multiplier $G = \frac{1}{3R^2} = \frac{1/3}{3600} = 0;0,0,0,20$,

$$Y_{k+1} = Y_0 + (Y_k)^3 \cdot G. \tag{5.6}$$

Verses 70–71/21–23. Assuming $X_0 = \frac{1}{3}R\sin\alpha$ and the multiplier $G = \frac{4}{3R^2} = \frac{4/3}{3600} = 0;0,0,1,20$,

$$X_{k+1} = X_0 + (X_k)^3 \cdot G. \tag{5.7}$$

5.2.3 Concluding Prose Material in the *Siddhāntakaustubha jyācāpa* Section

अथ प्रकारान्तरेण त्र्यंशजीवानयनं ।

Now the determination of the sine of three degrees by another method.

हरादनुमानेन फलं ग्राह्यं पूर्वपंक्तौ शून्यद्वयं स्थाप्यं । पुनः पुनः पंक्तिं फलगुणितां[18] हरे विकलास्थाने विशोध्य शेषं स्पष्टहरो भवति । उपरिशून्यद्वयस्य विद्यमानत्वाद्विकलास्थानमागतमेव । पुनर्लब्धिगुणं हरं विशोध्य[19] पंक्तिस्थफले फलं योजयित्वा फलेन पंक्तिर्गुणनीया । इयं यथास्थाने विशोध्य ततः पंक्तिस्थफले |[p. 1075 J_S] फलं पुनर्योज्यं । पंक्तिमस्तके एकं शून्यं निवेश्यं । ततो हरमुत्सार्य अनुमानेन फलं ग्राह्यं । फलं पंक्तौ निवेश्यं । ततः पंक्तिः फलगुणिता हरे विशोध्य शेषं स्पष्टहरो भवति । पुनर्लब्धगुणं हरं भाज्ये विशोध्य पंक्तिस्थफले फलं योजयित्वा फलेन पंक्तिर्गुणनिया । इयं हरे यथास्थाने विशोध्य ततो पंक्तिस्थफले फलं पुनर्योज्यं । पंक्त्यूर्ध्वे एकं शून्यं स्थापनीयं । ततो हरमुत्सार्यं अनुमानफलं ग्राह्यं । एवं मुहुः ।

The result obtained by proportion from the divisor is set in the first *paṅkti*, [and separately in a second *paṅkti*] with a pair of [leading] zeroes. Again [and] again, having subtracted the *paṅkti* multiplied by the result from the divisor in the seconds place, the remainder is the corrected divisor. Because of the presence of the pair of zeroes above, just the seconds place is attained. Again, having subtracted the quotient-multiplied divisor [from the dividend], [and] added the result to the *paṅkti*-placed result, the *paṅkti* is multiplied by the result. Subtracting this [from the corrected divisor] according to place, then the result again is added to the *paṅkti*-placed result. One zero is to be entered at the head of a [third] *paṅkti*. Then casting out the divisor, by proportion the [next] result is obtained. The result is to be entered in the [second] *paṅkti*. Then the [second] *paṅkti* is multiplied by the result subtracting [the product] from the divisor, the remainder is the corrected divisor. Again, subtracting the quotient-multiplied [corrected] divisor from the dividend,[and] adding the result to the *paṅkti*-placed result, the *paṅkti* is to be multiplied by the result. Subtracting this from the divisor according to place, then again the result is to be added to the *paṅkti*-placed result. At the top of the *paṅkti* one zero is to be placed. Then having cast out the [corrected] divisor, the proportional result is obtained. [And] so repeatedly.

अथोदाहरणं । भाज्यः ६ ।१६ ।४९ ।७ ।५९ ।८ ।५६ ।२९ ।४० हरः ३ हरेण ३ अनुमानात्फलं गृहीतं २ । पूर्वं पंक्तौ शून्यद्वयं स्थाप्यं ततः फलं स्थाप्यं ० ।० ।२ पुनः पंक्तिः फलगुणिता हरे विकलास्थाने[20] शोध्यशेषं स्पष्टहरो भवति २ ।५९ ।५६ अनेन भक्ते लब्धं २ फलगुणितं हरं भाज्ये विशोध्य शेषं ० ।१६ ।५७ ।७ ।५९ ।८ ।५६ ।२९ ।४० पुनः पंक्तिस्थफले फलं योजयित्वा फलेन पंक्तिर्गुणनीया । ८ इदं हरे यथास्थाने विशोध्य शेषं २ ।५९ ।४८ पंक्तिस्थफले फलं पुनर्योज्यं ।

Now, the example. Dividend: $6, 16, 49, 7, 59, 8, 56, 29, 40$. Divisor: 3. From proportion with the divisor 3 the result is obtained: 2. First, a pair of zeroes is set in the *paṅkti*, then the result is set: 0, 0, 2. Again, the *paṅkti* [is] multiplied by the result; the subtraction-remainder [from subtracting the product] from the divisor in the seconds place is the corrected divisor, $2, 59, 56$. When divided by this, the quotient [is] 2: subtracting the result-multiplied divisor from the dividend, the remainder: $0, 16, 57, 7, 59, 8, 56, 29, 40$. Again, having added the result to the *paṅkti*-placed result, the *paṅkti* is to be multiplied by the result. 8: subtracting this according to place from the divisor, remainder: $2, 59, 48$. The result is to be added to the *paṅkti*-placed result.

18 पंक्तिफलगुणितां (Sharma 1967–1969, p. 1074)

19 हरे विशोध्यं (Sharma 1967–1969, p. 1074)

20 विकला स्थाने (Sharma 1967–1969, p. 1075)

पंक्तिमस्तके एकं शून्यं निवेश्यं ०।०।०।६ ततो हरमुत्सार्य पुनरनुमानेन फलं ग्राह्यं ५ फलं पंक्तौ निवेश्यं ६।५ ततः पंक्तिः फलगुणिता[21] ३०।२५ हरे विशोध्य शेषं स्पष्टहरो भवति २।५९।४७।२९।३५ पुनर्लब्धगुणं हरं भाज्ये विशोध्य शेषं १।५८।१०।३१।१३ पंक्तिस्थफले $\genfrac{}{}{0pt}{}{\text{६}}{\text{५}}$ फलं ५ योजयित्वा $\genfrac{}{}{0pt}{}{\text{६}}{\text{१०}}$ फलेन पंक्तिर्गुणनीया $\genfrac{}{}{0pt}{}{\text{३०}}{\text{५०}}$[22] इदं हरे यथास्थाने विशोध्य |[p. 1076 J$_S$] शेषं २।५९।४६।५८।४५ पुनः पंक्तिस्थफले फलं योज्यं $\genfrac{}{}{0pt}{}{\text{६}}{\text{१५}}$ पंक्तिमस्तके एकं शून्यं स्थापनीयं । ततो हरमुत्सार्य अनुमानफलं ग्राह्यं शेषं पूर्ववत् । एवं मुहुः ।

One zero is to be entered at the head of the *paṅkti*: 0, 0, 0, 6. Then casting out the divisor, again by proportion the result is obtained: 5. The result is entered in the *paṅkti*: 6, 5. Then the *paṅkti* is multiplied by the result: 30, 25. Subtracting [that] from the divisor, the remainder is the corrected divisor: 2, 59, 47, 29, 35. Again, subtracting the quotient-multiplied divisor from the dividend, remainder: 1, 58, 10, 31, 13. Adding the result, 5, to the *paṅkti*-placed result, $\genfrac{}{}{0pt}{}{6}{5}$: $\genfrac{}{}{0pt}{}{6}{10}$. The *paṅkti* is multiplied by the result: $\genfrac{}{}{0pt}{}{30}{50}$. Subtracting this according to place from the divisor, remainder: 2, 59, 46, 58, 45. Again, the result is added to the *paṅkti*-placed result: $\genfrac{}{}{0pt}{}{6}{15}$. One zero is set at the head of the *paṅkti*. Then casting out the divisor, the proportional result is obtained; the rest is as before. [And] so repeatedly.

अत्रोपपत्तिः । श्रीमहाराजाधिराजजयसिंहदेवैर्निष्कासितास्ति सा यथा । सिद्धान्ते त्र्यंशज्यासाधनं । दोर्ज्यात्रिभागस्य[23] घनं तु तावत्कृत्वा हृतो ऽयं[24] त्रिगुणस्य कृत्या । स्वत्र्यंशयुक्तेन फलेन युक्तो दोर्ज्यात्रिभागः अनेन प्रकारेणासकृत् त्र्यंशज्यासाधनं कृतमस्ति । अत्र स्वत्र्यंशयुक्तफलेन[25] दोर्ज्यात्रिभागः युक्तो विधीयते । अथवा स्वत्र्यंशयुक्तफलं त्रिगुणं विधाय दोर्ज्यायां संयोज्य पश्चात्त्रिभिर्विभजनेनापि तुल्यमेव फलमायाति ।

Here is the demonstration: expounded by His Highness the supreme Mahārāja Jayasiṃha, it is as [follows]. In the *siddhānta* [is] the determination of the sine of three degrees (*try-aṃśa*). So having made the cube of a third part (*tri-bhāga*) of [that] sine [as] so much, that is divided by the square of the radius (*tri-guṇa*) [i.e., the sine of three (zodiacal signs)]. The third part (*tri-bhāga*) of the sine is added to the result increased by its own third part (*sva-try-aṃśa*): by this method iteratively the determination of the sine of the third part (*try-aṃśa*) [of the arc] is made. Here the third part (*tri-bhāga*) of the sine is considered increased by the result added to its own third part (*sva-try-aṃśa*). Or [equivalently], considering the result plus its own third part (*sva-try-aṃśa*) multiplied by three (*tri-guṇa*), [and] adding [it] to the sine of the arc, and after division by three, just the same result comes out.

अत्र तु ज्ञातदोर्ज्या पूर्णज्यात्वेन ज्ञातास्ति । तस्मात्स्वत्र्यंशयुक्तफलं[26] षड्गुणं विधाय पूर्णज्यायां संयोज्य त्रिभिर्विभाज्यं फलं त्र्यंशज्या भवति । षड्गुणस्य त्र्यंशफलं तु पूर्णज्यात्र्यंशस्य घनः वारद्वयं षष्ठिभक्त इति सिद्धं ।

But here the known sine is known by [virtue of] its being a chord. Because of that, considering the result plus its own third part (*sva-try-aṃśa*) multiplied by six (*ṣaḍ-guṇa*), [and] adding [it] to the chord [of 6°], the result divided by three is [twice] the sine of a third part (*try-aṃśa*) [of 3°]. Or, the third-part (*try-aṃśa*) result [being one-third] of the chord of six [degrees] (*ṣaḍ-guṇa*), [one-third] the cube of a third part (*try-aṃśa*) of the chord is twice divided by sixty: thus it is perfected.

---

21 पंक्तिफलगुणिता (Sharma 1967–1969, p. 1075)

22 $\genfrac{}{}{0pt}{}{\text{३}}{\text{५}}$ (Sharma 1967–1969, p. 1075)

23 दोर्ज्या त्रिभागस्य (Sharma 1967–1969, p. 1076)

24 हृतोयं (Sharma 1967–1969, p. 1076)

25 स्वत्र्यंशमुक्तफलेन (Sharma 1967–1969, p. 1076)

26 तस्मात्स्वत्र्यंशमुक्तफलं (Sharma 1967–1969, p. 1076)

5.2.4 Mathematical Notes on "Another Method" for Crd 2°

The final passage of this section commences with yet another method to determine the chord of two degrees (and consequently its half, $R \sin 1°$), through a highly condensed description of the general procedure followed by a similarly highly abridged worked numerical example. This method is identical to that of Dayānat Khān discussed in section 3.8, but applied to an equation for $y = \text{Crd}\, 2°$ rather than for $x = R \sin 1°$.

Given the parallels between the text's general description of the method and its following worked example, we present these two parts side by side, broken up into distinct stages of the iterative process. We also provide in italics additional steps of working and intermediate numerical results that have been omitted by the author but are crucial for reasoning. We hypothesize that, as with the exposition of Dayānat Khān's method in the *jyācāpa* manuscript recension (see section 2.8), the author of this passage likely envisioned the numerical steps of working to be placed in a grid, likewise with three *paṅkti* sections. We have reproduced what that might have looked like in figure 5.1.

Method / **Worked Example**

The result obtained by proportion from the divisor is set in the first *paṅkti*, [and separately in a second *paṅkti*] with a pair of [leading] zeroes.

Dividend: 6, 16, 49, 7, 59, 8, 56, 29, 40. Divisor: 3. From proportion with the divisor 3 the result is obtained: 2. First, a pair of zeroes is set in the *paṅkti*, then the result is set: 0, 0, 2.

The first step involves dividing the leading digit of the dividend, 6, *by the divisor,* 3. *The result,* 2, *is put both in the top row as* 2, *and at the head of the second paṅkti under the divisor with two leading zeros,* 0, 0, 2.

Again [and] again, having subtracted the *paṅkti* multiplied by the result from the divisor in the seconds place, the remainder is the corrected divisor.

Again, the *paṅkti* [is] multiplied by the result; the subtraction-remainder [from subtracting the product] from the divisor in the seconds place is the corrected divisor, 2, 59, 56.

This divisor in the second paṅkti, 0, 0, 2, *is multiplied by the result,* 2, *to produce* 0, 0, 4. *Then the divisor* 3, 0, 0 *is decreased by that product,* 0, 0, 4. *The result is the corrected divisor,* 2, 59, 56.

Because of the presence of the pair of zeroes above, just the seconds place is attained. Again, having subtracted the quotient-multiplied divisor [from the dividend]...

When divided by this, the quotient [is] 2: subtracting the result-multiplied divisor from the dividend, the remainder: 0, 16, 57, 7, 59, 8, 56, 29, 40.

Subtract the corrected divisor, 2, 59, 56 *multiplied by the quotient,* 2, *from the dividend; namely* 6, 16, 49, 7, 59, 8, 56, 29, 40 − 5, 59, 52. *The remainder is the new dividend,* 0, 16, 57, 7, 59, 8, 56, 29, 40.

... [and] added the result to the *paṅkti*-placed result, the *paṅkti* is multiplied by the result.

Again, having added the result to the *paṅkti*-placed result, the *paṅkti* is to be multiplied by the result. 8:...

The original "result", 2, *is added to the paṅkti-placed result, currently* 2, *and multiplied by the result in the second paṅkti,* 0, 0, 2. *The result is* 0, 0, 8.

Subtracting this [from the corrected divisor] according to place, then the result again is added to the *paṅkti*-placed result. One zero is to be entered at the head of a [third] *paṅkti*.

... subtracting this according to place from the divisor, remainder: 2, 59, 48. The result is to be added to the *paṅkti*-placed result. One zero is to be entered at the head of the *paṅkti*: 0, 0, 0, 6.

The corrected divisor, 2, 59, 56, *is decreased by this,* 0, 0, 8, *to produce* 2, 59, 48. *The result,* 0, 0, 2 *is added to the paṅkti-placed result,* 0, 0, 4. *The sum is* 0, 0, 6. *This amount is placed in the third paṅkti with another leading zero,* 0, 0, 0, 6.

Then casting out the divisor, by proportion the [next] result is obtained. The result is to be entered in the [second] *paṅkti*. Then the [second] *paṅkti* is multiplied by the result,

Then casting out the divisor, again by proportion the result is obtained: 5. The result is entered in the *paṅkti*: 6, 5. Then the *paṅkti* is multiplied by the result: 30, 25.

A new iteration begins. The leading digit of the new dividend, 16, *is now divided by* 3. *The result is* 5. *This is placed in the top row next to the* 2, *to obtain* 2, 5. *This is also placed in the second paṅkti next to the* 6, *to give* 6, 5. *This is multiplied by* 5 *to obtain the product* 30, 25. *This product is placed appropriately in the second paṅkti.*

. . .subtracting [the product] from the divisor, the remainder is the corrected divisor.

Subtracting [that] from the divisor, the remainder is the corrected divisor: 2, 59, 47, 29, 35.

This product, 30, 25, *is subtracted from the divisor,* 2, 59, 48, 0, 0 *to produce* 2, 59, 47, 29, 35. *This is the new corrected divisor.*

Again, subtracting the quotient-multiplied [corrected] divisor from the dividend,[and] adding the result to the *paṅkti*-placed result, the *paṅkti* is to be multiplied by the result.

Again, subtracting the quotient-multiplied divisor from the dividend, remainder: 1, 58, 10, 31, 13. Adding the result, 5, to the *paṅkti*-placed result, $\frac{6}{5}$: $\frac{6}{10}$. The *paṅkti* is multiplied by the result: $\frac{30}{50}$.

This quotient-multiplied divisor, 2, 59, 47, 29, 35×0, 5 = 0, 14, 58, 57, 27, 55 *is subtracted from the dividend in the first paṅkti,* 0, 16, 57, 7, 59, 8, 56, 29, 40, *to yield the remainder* 0, 1, 58, 10, 31, 13. *Adding the result,* 0, 5, *to the paṅkti-placed result,* 6, 5, *produces* 6, 10. *This is entered appropriately in the second paṅkti. This quantity is multiplied by the result,* 0, 5. *The product is* 30, 50.

Subtracting this from the divisor according to place, then again the result is to be added to the *paṅkti*-placed result. At the top of the *paṅkti* one zero is to be placed. Then having cast out the [corrected] divisor, the proportional result is obtained. [And] so repeatedly.

Subtracting this according to place from the divisor, remainder: 2, 59, 46, 58, 45. Again, the result is added to the *paṅkti*-placed result: $\frac{6}{15}$. One zero is set at the head of the *paṅkti*. Then casting out the divisor, the proportional result is obtained; the rest is as before. [And] so repeatedly.

Subtracting 0, 0, 0, 30, 50 *from the divisor,* 2, 59, 47, 29, 35 *yields* 2, 59, 46, 58, 45. *Then the result,* 0, 5, *is added to the paṅkti-placed result* 0, 0, 0, 6, 10 *to produce* 0, 0, 0, 6, 15. *The next iteration is invoked at this point by the phrase "casting out the divisor", but no further numerical data is given.*

	2	5							
1st	6	16	49	7	59	8	56	29	40
	5	59	52						
	0	16	57	7	59	8	56	29	40
		14	58	57	27	55			
		1	58	10	31	13			
2nd	3								
	0	0	2						
	0	0	4						
	2	59	56						
			8						
	2	59	48						
			6	*5*					
				30	25				
	2	59	47	29	35				
			6	*10*					
				30	50				
	2	59	46	58	45				
3rd	0	0	0	6					
				6	15				

Figure 5.1: A reconstructed computational grid for Jagannātha's "another method" for the chord of 2°, using the structure of the grid associated with Dayānat Khān's method in section 3.8.

More generally, the equations that generate updated values of the parameters for the next recursion cycle can be expressed as follows (compare their counterparts in section 3.8.1):

$$a_{k+1} = a_k - d_{k+1}(b_k + (0;0,1)d_{k+1}) - d_{k+1}(b_k + 2(0;0,1)d_{k+1}) \tag{5.8}$$
$$b_{k+1} = b_k + 2(0;0,1)d_{k+1} + (0;0,1)d_{k+1} \tag{5.9}$$
$$c_{k+1} = c_k - d_{k+1}(a_k - b_k d_{k+1} - (0;0,1)(d_{k+1})^2) \tag{5.10}$$
$$\tag{5.11}$$

The unknown value of y is broken into the sum of a known digit $d_0 = 2$ and an unknown remainder s_0. The "pre-initial" cycle, as we described in earlier in section 3.8, may here be considered to assign $a_{(-1)} = 3$, $b_{(-1)} = 0$, $c_{(-1)} = 6;16,49,7,59,8,56,29,40$, and $d_0 = 2$. The initial cycle with $k = 0$ then produces the following (we have highlighted in red those numbers which are recorded in the text's worked computation):

$$\begin{aligned} a_0 &= a_{(-1)} - d_0(b_{(-1)} + (0;0,1)d_0) - d_0(b_{(-1)} + 2(0;0,1)d_0) \\ &= 3 - 2(0;0,1)\cdot 2 - 2(2(0;0,1)\cdot 2) \\ &= 3 - 0;0,4 - 0;0,8 \\ &= 2;59,56 - 0;0,8 \\ &= 2;59,48 \end{aligned}$$

$$\begin{aligned} c_0 &= c_{(-1)} - d_0(a_{(-1)} - d_0(b_{(-1)} + (0;0,1)d_0)) \\ &= 6;16,49,7,59,8,56,29,40 - 2(2;59,56) \\ &= 0;16,57,7,59,8,56,29,40 \end{aligned}$$

$$\begin{aligned} b_0 &= b_{(-1)} + 2(0;0,1)d_0 + (0;0,1)d_0 \\ &= 0 + 0;0,4 + 0;0,2 \\ &= 0;0,6 \end{aligned}$$

The second iteration begins by setting $d_1 = 0;5$: Then

$$\begin{aligned} a_1 &= a_0 - d_1(b_0 + (0;0,1)d_1) - d_1(b_0 + 2(0;0,1)d_1) \\ &= 2;59,48 - (0;5)((0;0,6) + (0;0,1)(0;5)) - (0;5)((0;0,6) + 2(0;0,1)(0;5)) \\ &= 2;59,48 - (0;5)(0;0,6,5) - (0;5)(0;0,6,10) \\ &= 2;59,48 - 0;0,30,25 - 0;0,30,50 \\ &= 2;59,47,29,35 - 0;0,0,30,50 \\ &= 2;59,46,58,45 \end{aligned}$$

$$\begin{aligned} c_1 &= c_0 - d_1(a_0 - d_1(b_0 + (0;0,1)d_1)) \\ &= 0;16,57,7,59,8,56,29,40 - (0;5)(2;59,47,29,35) \\ &= 0;16,57,7,59,8,56,29,40 - 0;14,58,57,27,55 \\ &= 0;1,58,10,31,13 \end{aligned}$$

$$\begin{aligned} b_1 &= b_0 + 2(0;0,1)d_1 + (0;0,1)d_1 \\ &= (0;0,6 + 0;0,0,10) + 0;0,0,5 \\ &= 0;0,6,10 + 0;0,0,5 \\ &= 0;0,6,15 \end{aligned}$$

Although at this point $d_2 = 0;0,39$ could be determined, the author stops.

5.2.5 Mathematical Notes on the Concluding Jayasiṃha Exposition

The final exposition, explicitly attributed to Mahārāja Jayasiṃha himself, is ingeniously (not to say maddeningly) embellished with playfully ambiguous technical terms. Nonetheless, it is evident that he is describing the successive-values iterations explained in section 5.2.2.2. The main concepts for which double-meaning terms are employed are the following:

Part/Degree Sanskrit *aṃśa* and *bhāga* can both mean either an arbitrary "part", as in the denominator of a fraction, or more specifically a "degree", $1/360$ of a circle.

Sine/Chord/Multiple A synonym for the usual Sanskrit terms for sine, *jyā* and *jīvā*, is *guṇa*, literally "string": all three terms connote the bow-and-bowstring configuration of the arc and chord. But *guṇa* in mathematics also means a multiplier or coefficient. (To avoid confusion between chord and sine, the chord is frequently designated *pūrṇa-jyā* "full chord".)

He begins with the algorithm for $x = R \sin 1°$ stated in verse *SK* II.1.67/J$_S$ 19, taking the angle α to be "three degrees" (*try-aṃśa*). Its "sine" (*jyā*), as he mentions, is known from the literature. A "third part" (*tri-bhāga*) of that known value is cubed and divided by the square of the trigonometric "radius" (*tri-guṇa*). That quotient plus "its own third part" (*sva-try-aṃśa*) is added to the "third part" (*tri-bhāga*) of the original sine, and the procedure is iterated to produce the sine of the "third part" (*try-aṃśa*) of the original arc. In other words, assuming $x_0 = \frac{1}{3} R \sin 3°$,

$$x_{k+1} = x_0 + \frac{(x_k)^3}{R^2} \cdot \left(1 + \frac{1}{3}\right), \quad k \in \{0, 1, 2, 3, \ldots\}. \tag{5.12}$$

Equivalently, one could take the quotient plus "its own third part" (*sva-try-aṃśa*), "multiplied by three" (*tri-guṇa*), and add it to the entire known sine value, subsequently dividing that sum by three:

$$x_{k+1} = \left(R \sin 3° + \frac{(x_k)^3}{R^2} \cdot \left(1 + \frac{1}{3}\right) \cdot 3\right) \div 3. \tag{5.13}$$

Equally equivalently, the chord of 2° or twice the sine of 1° can be generated by the same quotient plus "its own third part" (*sva-try-aṃśa*), "multiplied by six" (*ṣaḍ-guṇa*), added to the "chord" (*pūrṇa-jyā*) of 6°, and then divided by three:

$$\begin{aligned} 2\,x_{k+1} &= \left(2\,R \sin 3° + \frac{(x_k)^3}{R^2} \cdot \left(1 + \frac{1}{3}\right) \cdot 6\right) \div 3 \\ &= \left(\mathrm{Crd}\, 6° + \frac{(x_k)^3}{R^2} \cdot \left(1 + \frac{1}{3}\right) \cdot 6\right) \div 3. \end{aligned} \tag{5.14}$$

The final sentence is somewhat confusing but seems to be a paraphrase of the algorithm for $y = \mathrm{Crd}\, 2°$ in verses *SK* II.1.68–69/J$_S$ 20–21, stipulating $R = 60$. Here it is the "chord of six" degrees (*ṣaḍ-guṇa*) that is divided by three, and then presumably added to the quotient from dividing one-third of its own cube by the square of 60:

$$\text{assuming} \quad R = 60 \quad \text{and} \quad y_0 = \frac{1}{3}\mathrm{Crd}\, 6°, \qquad y_1 = y_0 + \frac{(y_0)^3}{3} \div 3600,$$
$$y_{k+1} = y_0 + \frac{(y_k)^3}{3} \div 3600. \tag{5.15}$$

Chapter 6
Reflections on the Three Techniques of $R \sin 1°$ Iteration

The profusion of iterative algorithms for finding various forms of $R \sin 1°$ and Crd $2°$ that we have examined in this study testifies to the interest the problem inspired among early modern trigonometers writing in Arabic, Persian and Sanskrit. But the specific interconnections of these closely related algorithms still remain quite obscure: the questions they raise are too numerous to list, much less answer fully in the present work. In the remaining pages we briefly summarize some of the prospects for further investigation.

Structure and History of the Methods

We have classified the three main types of approach to the problem as follows:

- The successive-digits technique, or digit-recurrence algorithm, discussed in section A.2. Ascribed to al-Kāshī in various sources (Rosenfeld and Hogendijk 2003), it was apparently developed not very long before his death in 1429 (Van Brummelen n.d.). Its operation for $y = \text{Crd}\, 2°$ has been analyzed in sections 3.3 and 4.3.3.
- The distinctive form of successive-digits iteration attributed to Dayānat Khān sometime around 1730, and explained for $x = R \sin 1°$ in section 3.8 (with an analogous y-version treated in section 5.2.4). This is the most operationally efficient of the three methods, in terms of having to carry out fewer arithmetic operations to obtain the same precision as in al-Kāshī's original, somewhat less efficient technique.
- The successive-values iteration apparently originating with al-Kāshī's contemporary and employer Ulugh Beg (according to al-Bīrjandī's attribution about a hundred years afterward, quoted in section 4.2). It is discussed for $x = R \sin 1°$ in sections 3.6 and 4.3.4, while its y-counterparts are treated in sections 3.4, 3.5, and 4.3.4. It is conceptually simpler, but less efficient, than the successive-digits recursions.

In the following schematic, we illustrate these variations by comparing the basic structure of three different types of algorithms for $x = R \sin 1°$ with $R = 60$, accompanied by partially worked examples from the *jyācāpa* excerpt. (Note that, as indicated in table 3.1, our Sanskrit manuscript of the *jyācāpa* material does not present any Kāshī-type successive-digits method specifically for $x = R \sin 1°$; we have relied instead on the description of such a method in the Arabic treatise presented in (Rosenfeld and Hogendijk 2003).)

C. Montelle et al., *From Samarqand to Jaipur*, Sources and Studies in the History of Mathematics and Physical Sciences, https://doi.org/10.1007/978-3-031-89087-1_6

Al-Kāshī's successive-digits (§ A.2)	Dayānat Khān's successive-digits (§ 3.8)	Successive-values (§§ 3.6 and 4.3.4)
Underlying equation(s)		
Specified as (Rosenfeld and Hogendijk 2003, p. 46): "45 0 *things* are equal to a *cube* and 47 6 8 29 53 37 3 37 15 sevenths", equivalent to the cubic equation $(45,0)x = x^3 + 47,6;8,29,53,37,3,37,15 \approx x^3 + (15,0)R\sin 3°.$	Specified as (§ 2.8): *yā* 3 *yāgha* 4″ *rū* 3; 8, 24, 33, 59, 34, 28, 15 equivalent to the cubic equation $3x = (0;0,4)x^3 + 3;8,24,33,59,34,28,15 \approx \frac{x^3}{15,0} + R\sin 3°,$ or $3;8,24,33,59,34,28,15 = x\left(3-(0;0,4)x^2\right).$	Specified as (§ 2.6): *yā* 1 *yāgha* 1″, 20‴ *rū* 1; 2, 48, 11, 19, 51, 29, 25 equivalent to the cubic equation $x = (0;0,1,20)x^3 + 1;2,48,11,19,51,29,25 \approx \frac{x^3}{45,0} + \frac{R\sin 3°}{3}.$
Fundamental technique		
Digit recurrence. As in the analogous method for y (see § 3.3.1), the value of x in the kth iteration cycle is expressed as a sequence of k sexagesimal digits: $x_k = q_0;q_1,q_2,\ldots,q_k$ Dividing the constant term by 45, 0 produces an initial estimate of x. In each kth cycle, the remainder r_{k-1} from the previous cycle's integer division by 45, 0 is added to the difference of the cubes of the two previous x-values. (In the very first cycle, where there is only one cube $(x_0)^3$, the subtracted cube can be considered zero.) Integer division then yields the next digit q_k, suitably adjusted for place-value: $\ldots,q_k,\ldots = \frac{r_{k-1} + (x_{k-1})^3 - (x_{k-2})^3}{45,0}$ Less significant digits are sometimes discarded without affecting the result.	Digit recurrence. x is treated recursively as the sum of a known sexagesimal digit d_0 plus an unknown remainder s_0, which itself is the sum of a digit and a remainder: $x = d_0 + s_0 = d_0 + d_1 + s_1 \ldots = d_0 + d_1 + \ldots + d_k + s_k$ Rewriting the underlying equation as $3;8,24,33,59,34,28,15 = (d_0+s_0)\left(3-(0;0,4)(d_0+s_0)^2\right),$ in each kth recursion cycle, an expression of this form is expanded to permit separating the known d terms from unknown s terms, yielding a modified equation of the form $c_k = (d_{k+1}+s_{k+1})[\,a_k - b_k(d_{k+1}+s_{k+1}) - (0;0,4)(d_{k+1}+s_{k+1})^2\,]$ The new digit d_{k+1}, found by inspection, determines the next coefficients a_{k+1}, b_{k+1}, and c_{k+1}.	Fixed-point iteration. The equivalent iterative equation, for $k \in \{0,1,2,\ldots\}$, is $x_{k+1} = x_0 + (0;0,1,20)\,(x_k)^3$

Al-Kāshī's successive-digits (§ A.2)	Khān's successive-digits (§ 3.8)	Successive-values (§ 3.6 and 4.3.4)
	Initial value	
$q_0 = (47,6;8,29,53,37,3,37,15)$ $\div (45,0)$ $= 1$	$d_0 = (3;8,24,33,59,34,28,15) \div 3$ $= 1$	$x_0 = 1;2,48,11,19,51,29,25$
	Iteration cycle 1	
$r_0 = (47,6;8,29,53,37,3,37,15)$ $- (45,0)\, q_0$ $= \quad 47,6;8,29,53,37,3,37,15$ $-45,0$ $\quad 2,6;8,29,53,37,3,37,15$ This remainder is then added to the cube of the first approximation $x_0 = q_0$, and the sum is divided to yield the next digit q_1: $r_0 + (x_0)^3 = \quad 2,6;8,29,53,\ldots$ $+ \quad 1$ $\quad 2,7;8,29,53,\ldots$ $0;q_1 = \left(r_0 + (x_0)^3\right) \div (45,0)$ $= (2,7;8,29,53,\ldots) \div (45,0)$ $= 0;2$ $x_1 = q_0;q_1 = 1\ 2$	(For further details, see § 3.8.2.) Using "pre-initial" coefficients from the original cubic equation: $a_{(-1)} = 3, \quad b_{(-1)} = 0,$ $c_{(-1)} = 3;8,24,33,59,34,28,15,$ compute by the prescribed operations $a_0 = 2;59,48, \quad b_0 = 0;0,12$ $c_0 = 0;8,28,33,59,34,28,15.$ Find by inspection the largest d_1 such that $d_1 \cdot a_0 \leq c_0$, namely, $d_1 = 0;2.$ $x \approx d_0 + d_1 = 1;2$	(For further details, see § 3.6.2.) $x_1 = x_0 + (0;0,1,20)\,(x_0)^3$ $= \quad 1;2,48,11,19,51,29,25$ $+0;0,\ 1,31,44,40,31,28,41$ $\quad 1;2,49,43,\ 4,32,\ 0,53,41$
	Iteration cycle 2	
$r_1 = (2,7;8,29,53,\ldots)$ $-(45,0)(0;q_1)$ $= \quad 2,\ 7;8,29,53,\ldots$ $-1,30$ $\quad 37;8,29,53,\ldots$ $r_1 + (x_1)^3 - (x_0)^3$ $= \quad 37;\ 8,29,53,\ldots$ $+ \quad 1;\ 6,12,\ 8$ $- \quad 1$ $\quad 37;14,42,\ 1,\ldots$ $0;0,q_2 = \left(r_1 + (x_1)^3 - (x_0)^3\right)$ $\div (45,0)$ $= (37;14,42,1,\ldots) \div (45,0)$ $= 0;0,49$ $x_2 = q_0;q_1,q_2 = 1;2,49$	Using a_0, b_0, c_0, d_1, compute by the prescribed operations $a_1 = 2;59,47,11,12,$ $b_1 = 0;0,12,24,$ $c_1 = 0;2,28,58,48,6,28,15.$ Find by inspection the largest d_2 such that $d_2 \cdot a_1 \leq c_1$, namely, $d_2 = 0;0,49.$ $x \approx d_0 + d_1 + d_2 = 1;2,49$	$x_2 = x_0 + (0;0,1,20)\,(x_1)^3$ $= \quad 1;2,48,11,19,51,29,25$ $+0;0,\ 1,31,51,22,45,25,8$ $\quad 1;2,49,43,11,14,14,50,8$

Al-Kāshī's original successive-digits iterative approximation was apparently initially inspired by the familiar digit-recurrence algorithm for place-value long division, extracting the desired quotient one digit at a time, that he explained in his *Miftāḥ al-Ḥisāb* or *Key of Arithmetic* (Aydin and Hammoudi 2019, pp. 61–75). In this context, it is perhaps significant that al-Bīrjandī at the end of his description of al-Kāshī's method for Crd 2° (section 4.2) refers to the final approximate result in his worked-example grid as a "quotient" (*qismat*). The Arabic treatise likewise speaks of al-Kāshī's introducing the cube term "into the division" (*qasm*) (Rosenfeld and Hogendijk 2003, pp. 42, 58), commencing with the division of an equation such as equation 1.4 by the coefficient of the unknown.

Dayānat Khān's successive-digits approximation takes a somewhat different approach to digit recurrence, one that fits within a class of solutions to polynomial equations attested in several cultures. Al-Kāshī himself addressed root-finding procedures, including the extraction of a fifth root, in a similar way in the *Miftāḥ al-Ḥisāb* (Aydin and Hammoudi 2019, pp. 75–115), (Aydin and Hammoudi 2015), (Berggren 1986, pp. 53–63). Related methods for calculating square and cube roots are found among al-Kāshī's predecessors dating back several hundred years, occasionally attributed to Indian sources. Indeed, such methods for place-value square- and cube-root extraction appear in Sanskrit texts as early as the fifth century CE. This approach also occurs in pre-modern Chinese solutions for roots, not only of integer powers of a given number, but sometimes other polynomials as well. All such methods work recursively, beginning by determining the leading digit of the solution of the given polynomial (more or less as in long division). Then the original unknown is replaced in the polynomial with the sum of that digit and an unknown remainder. The result is expanded, giving a new polynomial whose unknown is an order of magnitude smaller than the original. The method is then repeated again and again, determining one digit with each iteration.

In the fixed-point iteration of successive-values algorithms, on the other hand, the iterative cubic equation is solved in each cycle for the entire multi-digit unknown, converging to a progressively more accurate value. As we mentioned in section 1.1.2, such fixed-point approximations had been part of the standard toolkit of Indian applied trigonometry for centuries, although the specific algorithm ascribed by al-Bīrjandī to Ulugh Beg (and subsequently explored in many variant forms by authors including Nityānanda and Jagannātha) seems to have originated in Samarqand.

Computational Efficiency and Diversity

It is clear that the greater procedural simplicity of the successive-values or fixed-point iteration approach does not translate into greater operational efficiency. It is easier to theoretically prescribe the repeated cubing of multi-digit sexagesimal numbers than to carry it out in practice! Al-Kāshī's successive-digits approach, although more cumbersome to explain, is more efficient in its restriction of sexagesimal cubing to the comparatively short digit sequences beginning with 1 and 1; 2.

In Dayānat Khān's digit-recurrence method, the recursion has been cleverly modified to minimize division or cubing of multiple sexagesimal digits. Procedurally, however, it is extremely opaque. The complicated numerical grid used to carry out the prescribed operations appears far removed from the original cubic equation that gave rise to it; as demonstrated in section 3.8, a great deal of elaborate algebraic manipulation is required to fully understand the connection between them.

Each of these three distinct types of approximation method, as we have discussed, covers multiple more or less trivial variants of the basic algorithm. The reader may well wonder why so much effort was devoted to devising supplementary and very similar methods for calculating a reasonably accurate approximate value of a standard irrational constant such as $R \sin 1^\circ$. It may seem unnecessary to re-derive a value in a slightly different way, to the same level of accuracy, after it already had been computed and made available for use in trigonometric tables. Clearly the problem captured the attention of the courts at Samarqand and at Jaipur, stimulating curiosity that explored beyond al-Kāshī's original solution. The importance of the problem and al-Kāshī's introduction of a novel form of solution that bypassed its apparent theoretical impossibility may have been sufficient to ensure the continuing interest taken in them by generations of mathematicians.

Continuing Exploration of $R \sin 1^\circ$ Iterations

Our study of the *jyācāpa* excerpt and related material has revealed at every step how much remains to be learned about the roles of numerical approximation in pre-modern mathematics. We will confine ourselves in these closing remarks to surveying a few relevant subjects for forthcoming exploration:

- *Detailed analysis of the mechanics and operational efficiency of various sine iterations*
  It seems reasonable that the goal of reducing computational labor of computation played at least some part in the creation of different types of $R \sin 1^\circ$ iterations. A detailed reconstruction of the operational processes involved in each of the methods discussed here, and a quantitative comparison of their computing performance, should shed more light on this issue.
- *Possible related $R \sin 1^\circ$ methods in contemporary texts*
  In this study we have been able to trace some of the direct scholarly lineage of the algorithms in the *jyācāpa* text. But we do not yet know what other authors might have worked on this problem, or how they might have influenced the evolution of these techniques.

- *Non-technical incentives for developing variant methods*
Sources such as Jamshīd al-Kāshī's own letters from the Samarqand court make it clear that impressing one's colleagues and patrons with one's mathematical knowledge and problem-solving skills was desirable for professional success. There are somewhat similar allusions in Sanskrit sources to the importance of demonstrating ability in "assemblies of the learned". Presumably a mathematical achievement like being able to determine an accurate value of $R \sin 1°$, rather than merely computing with a previously known value, was professionally valuable in Nityānanda's Delhi and Nayanasukha's Jaipur as well as in al-Kāshī's Samarqand. But the social and intellectual history of such phenomena in pre-modern science is still very incomplete.
- *Scientific communities and communication in the early modern Islamic and Indian worlds*
As the foregoing discussions have shown, early modern interest in the topic of sine $1°$ approximations was wide-ranging both geographically and linguistically. We would like to know more about, among others, the following aspects of these developments:
  - *The role of the astronomer-princes*
  In 18th-century Jaipur as in 15th-century Samarqand, the pursuit of mathematical research was shared, as well as sponsored, by the ruler. How did this affect the conduct of the research, beyond the obvious advantages of a favorable funding environment?
  - *Translation and collaboration*
  How did colleagues in the Jaipur court work together across the linguistic boundaries of different text corpora? What was the role of vernacular interpreters in these discussions, as compared to direct study of a collaborator's language?
- *The shaping of a Sanskrit mathematical topic*
The *jyācāpa* material's detailed discussion of $R \sin 1°$ iterative approximations and their sources offers a fascinating and rare example of the process of establishing a particular problem within the Sanskrit mathematical corpus. Most Sanskrit mathematical works let the content and style of technical exposition speak for themselves, with little in the way of historical background or professional context. But in the *jyācāpa* material we see Sanskrit authors directly quoting, critiquing, and otherwise engaging with existing works, while refining the material into traditional Sanskrit didactic form for, presumably, a traditional pandit readership. The process of transforming the rhetoric of Islamic algebra demonstrations into Sanskrit counterparts, for example, is intriguingly hinted at in some of the small differences in the text of MSS J_1 and J_2. We hope that continuing study along the lines sketched out above will reveal more of the structure of this process.

Appendix A
Calculations for Successive-Digits Iterations in the Anonymous Arabic Treatise

This appendix lays out calculations that underscore the numerical data produced from performing the prescribed iterative successive-digits procedure for the chord of 2° and the sine of 1° respectively in the anonymous Arabic treatise. These data are displayed in two separate tables in the primary source (Rosenfeld and Hogendijk 2003, p. 43, 58 and p. 47, 60). See figures A.1 and A.2 and tables A.1 and A.2 for facsimiles and transcriptions of these.

A.1 Successive-Digits Iteration for the Chord of 2°

This successive-digits iteration procedure for the chord of 2° is based on the following cubic equation:

$$y = \frac{6, 16, 49; 7, 59, 8, 56, 29, 40 + y^3}{3, 0, 0}.$$

The steps of the numerical procedure are not spelled out in the text; instead the numerical data is presented *fait accompli* in a table, introduced by "This is a table of the operation" (Rosenfeld and Hogendijk 2003, pp. 42–43). The reconstructed computational steps are presented over four columns below for each of the nine steps of the iterative procedure. Column four documents how the numerical data is put together step by step in the grid recorded in table A.1. Given the fact that these numerical values are aligned in a grid, thus keeping track of the absolute value of the numbers relative to one another, for simplicity we present our analysis in floating point notation.

In a manner similar to the operations set out in section 3.3, the iterative procedure generates a "result-row" or current approximate value $y \approx q_1, q_2, \ldots, q_k$ that is extracted digit by digit,[1] with each digit q_i and corresponding "dividend" term p_i and "remainder" term r_i computed from the ones used in the preceding iteration.

[1] In this particular case, the first two digits of the result-row are generated in one step, hence we begin with q_1.

C. Montelle et al., *From Samarqand to Jaipur*, Sources and Studies in the History of Mathematics and Physical Sciences, https://doi.org/10.1007/978-3-031-89087-1

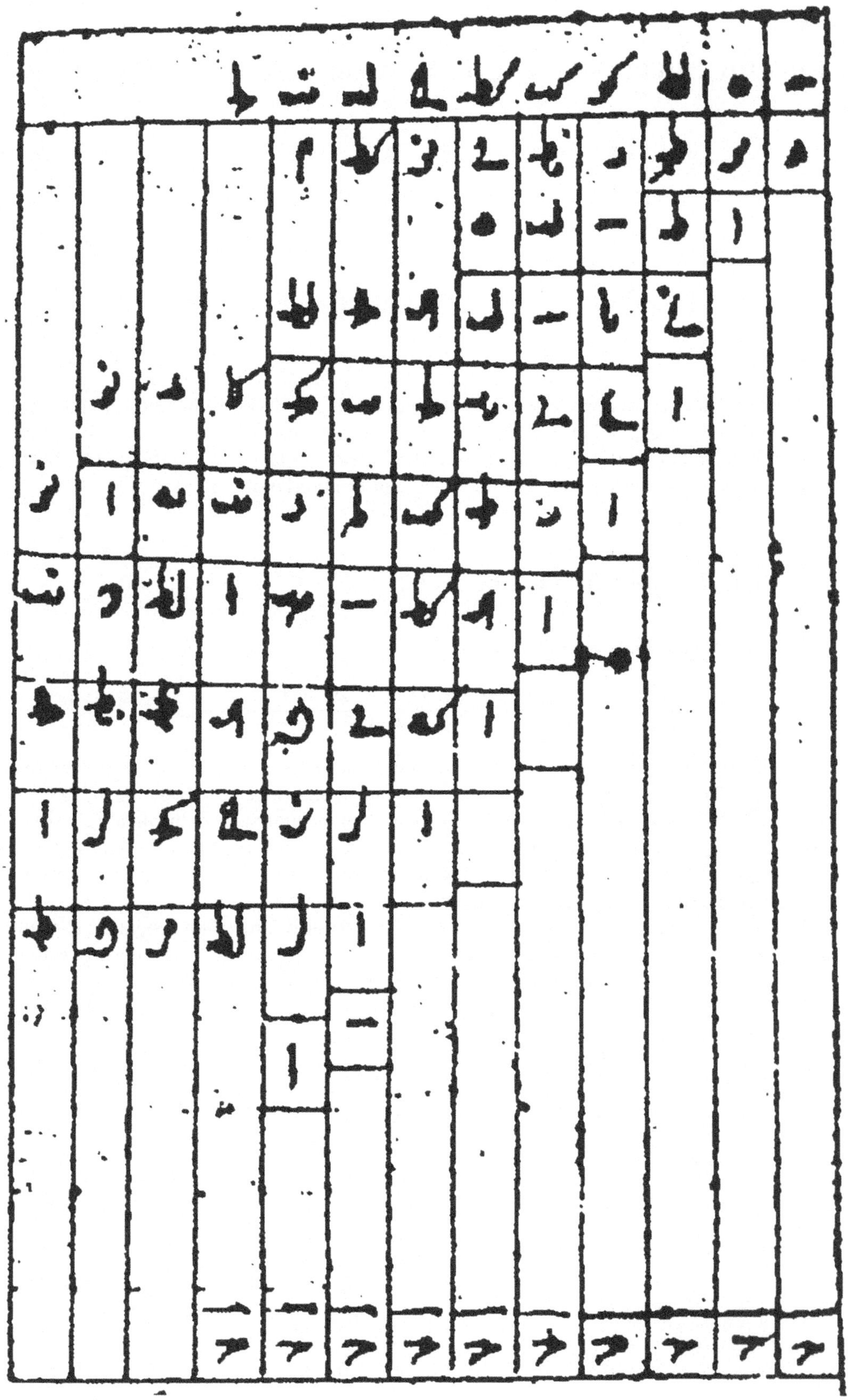

Figure A.1: Facsimile of the Tehran lithograph edition (j; p. 45) of the anonymous Arabic treatise (Rosenfeld and Hogendijk 2003, p. 58), showing the numerical data produced from performing the prescribed iterative procedure for the chord of 2°. The value for the chord of 2° (2; 5, 39, 26, 22, 29, 28, 32, 52, 33) appears along the top row from right to left in Arabic *abjad* notation. This grid is reproduced in table A.1.

2	5	39	26	22	29	28	32	52	33			
6	16	49	7	59	8	56	29	40				
	1	9	2	32	5							
		58	11	2	32	27	43	39				
		1	18	8	14	33	12	23	21	4	56	
			1	7	19	22	41	6	52	15	1	56
				1	27	29	2	42	1	39	50	52
					1	25	8	50	27	19	59	43
						1	37	57	28	23	37	1
							1	37	39	47	50	49
							2	1				
3	3	3	3	3	3	3	3	3	3			

Table A.1: Translation of the numerical grid in figure A.1.

k	p_k	q_k, r_k, and result-row $q_0, q_1, q_2, \ldots, q_k$
1	Crd 6° $= 6, 16, 49, 7, 59, 8, 56, 29, 40$ $p_1 = 6, 16$	$q_1 = p_1 \div 3$ $= 6, 16 \div 3 \quad = 2, 5$ $r_1 = p_1 - 3\,q_1$ $= 6, 16 - 6, 15$ $= 0, 1$ result-row: 2, 5
2	$(q_1)^3 = (2, 5)^3 \quad = 9, 2, 32, 5$ $r_1^* = r_1 +$ third sexagesimal digit of Crd 6° $= 0, 1 + 0, 0, 49$ r_1^* : 0, 1, 49 + 9 $p_2 =$ 1, 58	$q_2 = p_2 \div 3$ $= 1, 58 \div 3 \quad = 39$ $r_2 = p_2 - 3\,q_2$ $= 1, 58 - 1, 57$ $= 0, 1$ result-row: 2, 5, 39
3	$(q_1, q_2)^3 = (2, 5, 39)^3$ $= 9, 11, 2, 32, 27, 43, 39$ $r_2^* = r_2 +$ fourth sexagesimal digit of Crd 6° $= 0, 1 + 0, 0, 7$ r_2^* : 0, 1, 7 + 11 $p_3 =$ 1, 18	$q_3 = p_3 \div 3$ $= 1, 18 \div 3 \quad = 26$ $r_3 = p_3 - 3\,q_3$ $= 1, 18 - 1, 18$ $= 0, 0$ result-row: 2, 5, 39, 26
4	$(q_1, q_2, q_3)^3 = (2, 5, 39, 26)^3$ $= 9, 11, 8, 14, 33, 12, 23, 21, 4, 56$ $r_3^* = r_3 +$ fifth sexagesimal digit of Crd 6° $= 0, 0 + 0, 0, 59$ r_2^* : 0, 0, 59 + 8 $p_4 =$ 1, 7	$q_4 = p_4 \div 3$ $= 1, 7 \div 3 \quad = 22$ $r_4 = p_4 - 3\,q_4$ $= 1, 7 - 1, 6$ $= 0, 1$ result-row: 2, 5, 39, 26, 22

Record Crd 6°.

Place $q_1 = 2, 5$ above in the top row of the grid (the result-row).

2	**5**							
6	16	49	7	59	8	56	29	40

Record $(q_1)^3 = 9, 2, 32, 5$ underneath.

Record $p_2 = 1, 58$.

Add $q_2 = 39$ to the top result-row.

2	5	**39**						
6	16	49	7	59	8	56	29	40
		9	**2**	**32**	**5**			
	1	**58**						

Record $(q_1, q_2)^3 = 9, 11, 2, 32, 27, 43, 39$ beginning from the second digit.

Record $p_3 = 1, 18$.

Add $q_3 = 26$ to the top result-row.

2	5	39	**26**					
6	16	49	7	59	8	56	29	40
		9	2	32	5			
	1	58	**11**	**2**	**32**	**27**	**43**	**39**
		1	**18**					

Record $(q_1, q_2, q_3)^3$ = 9, 11, 8, 14, 33, 12, 23, 21, 4, 56 beginning from the third digit.

Record $p_4 = 1, 7$.

Add $q_4 = 22$ to the top result-row.

2	5	39	26	**22**							
6	16	49	7	59	8	56	29	40			
		9	2	32	5						
	1	58	11	2	32	27	43	39			
		1	18	**8**	**14**	**33**	**12**	**23**	**21**	**4**	**56**
			1	**7**							

k	p_k	$q_k,\ r_k,$ and result-row $q_0, q_1, q_2, \ldots, q_k$
5	$(q_1, q_2, q_3, q_4)^3 = (2, 5, 39, 26, 22)^3$ $= 9, 11, 8, 19, 22, 41, 6, 52, 15, 1, 56$ $r_4^* = r_4 +$ sixth sexagesimal digit of Crd 6° $= 1 + 0, 8$ r_4^*: 0, 1, 8 + 19 $p_5 =$ 1, 27	$q_5 = p_5 \div 3$ $= 1, 27 \div 3 \quad = 29$ $r_5 = p_5 - 3\,q_5$ $= 1, 27 - 1, 27$ $= 0, 0$ result-row: 2, 5, 39, 26, 22, 29
6	$(q_1, q_2, q_3, q_4, q_5)^3 = (2, 5, 39, 26, 22, 29)^3$ $= 9, 11, 8, 19, 29, 2, 42, 1, 39, 50, 52$ $r_5^* = r_5 +$ seventh sexagesimal digit of Crd 6° $= 0, 0 + 0, 56$ r_5^*: 0, 56 + 29 $p_6 =$ 1, 25	$q_6 = p_6 \div 3$ $= 1, 25 \div 3 \quad = 28$ $r_6 = p_6 - 3\,q_6$ $= 1, 25 - 1, 24$ $= 0, 1$ result-row: 2, 5, 39, 26, 22, 29, 28
7	$(q_1, q_2, q_3, q_4, q_5, q_6)^3 = (2, 5, 39, 26, 22, 29, 28)^3$ $= 9, 11, 8, 19, 29, 8, 50, 27, 19, 59, 43$ $r_6^* = r_6 +$ eighth sexagesimal digit of Crd 6° $= 0, 1 + 0, 0, 29$ r_6^*: 0, 1, 29 + 8 $p_7 =$ 1, 37	$q_7 = p_7 \div 3$ $= 1, 37 \div 3 \quad = 32$ $r_7 = p_7 - 3\,q_7$ $= 1, 37 - 1, 36$ $= 0, 1$ result-row: 2, 5, 39, 26, 22, 29, 28, 32

Record
$(q_1, q_2, q_3, q_4)^3$
$= 9, 11, 8, 19, 22, 41, 6, 52, 15, 1, 56$
beginning from the fourth digit.

Record $p_5 = 1, 27$.

Add $q_5 = 29$ to the top result-row.

2	5	39	26	22	**29**							
6	16	49	7	59	8	56	29	40				
		9	2	32	5							
	1	58	11	2	32	27	43	39				
		1	18	8	14	33	12	23	21	4	56	
			1	7	**19**	**22**	**41**	**6**	**52**	**15**	**1**	**56**
				1	**27**							

Record $(q_1, q_2, q_3, q_4, q_5)^3$
$= 9, 11, 8, 19, 29, 2, 42, 1, 39, 50, 52$
beginning from the fifth digit.

Record $p_6 = 1, 25$.

Add $q_6 = 28$ to the top result-row.

2	5	39	26	22	29	**28**						
6	16	49	7	59	8	56	29	40				
		9	2	32	5							
	1	58	11	2	32	27	43	39				
		1	18	8	14	33	12	23	21	4	56	
			1	7	19	22	41	6	52	15	1	56
				1	27	**29**	**2**	**42**	**1**	**39**	**50**	**52**
					1	**25**						

Record
$(q_1, q_2, q_3, q_4, q_5, q_6)^3$
$= 9, 11, 8, 19, 29, 8, 50, 27, 19, 59, 43$
beginning from the sixth digit.

Record $p_7 = 1, 37$.

Add $q_7 = 32$ to the top result-row.

2	5	39	26	22	29	28	**32**					
6	16	49	7	59	8	56	29	40				
		9	2	32	5							
	1	58	11	2	32	27	43	39				
		1	18	8	14	33	12	23	21	4	56	
			1	7	19	22	41	6	52	15	1	56
				1	27	29	2	42	1	39	50	52
					1	25	**8**	**50**	**27**	**19**	**59**	**43**
						1	**37**					

k	p_k	q_k, r_k, and result-row $q_0, q_1, q_2, \ldots, q_k$
8	$(q_1, q_2, q_3, q_4, q_5, q_6, q_7)^3 = (2, 5, 39, 26, 22, 29, 28, 32)^3$ $= 9, 11, 8, 19, 29, 8, 57, 28, 23, 37, 1$ $r_7^* = r_7$ + ninth sexagesimal digit of Crd 6° $= 1 + 0, 40$ r_7^* : 0, 1, 40 + 57 $p_8 =$ 2, 37	$q_8 = p_8 \div 3$ $= 2, 37 \div 3 \quad = 52$ $r_8 = p_8 - 3\,q_8$ $= 2, 37 - 2, 36$ $= 0, 1$ result-row: 2, 5, 39, 26, 22, 29, 28, 32, 52
9	$(q_1, q_2, q_3, q_4, q_5, q_6, q_7, q_8)^3 = (2, 5, 39, 26, 22, 29, 28, 32, 52)^3$ $= 9, 11, 8, 19, 29, 8, 57, 39, 47, 50, 19$ $r_8^* = r_8$ + tenth sexagesimal digit of Crd 6° $= 0, 1 +$ [0, 0] r_8^* : 1, 0 + 39 $p_9 =$ 1, 39	$q_9 = p_9 \div 3$ $= 1, 39 \div 3 \quad = 33$ $r_9 = p_9 - 3\,q_9$ $= 1, 39 - 1, 39$ $= 0, 0$ result-row: 2, 5, 39, 26, 22, 29, 28, 32, 52, 33

Record $(q_1, q_2, q_3, q_4, q_5, q_6, q_7)^3$
$= 9, 11, 8, 19, 29, 8, 57, 28, 23, 37, 1$
beginning from the seventh digit.

Record $p_8 = 2, 37$.

Add $q_8 = 52$ to the top result-row.

2	5	39	26	22	29	28	32	**52**				
6	16	49	7	59	8	56	29	40				
		9	2	32	5							
	1	58	11	2	32	27	43	39				
		1	18	8	14	33	12	23	21	4	56	
			1	7	19	22	41	6	52	15	1	56
				1	27	29	2	42	1	39	50	52
					1	25	8	50	27	19	59	43
						1	37	**57**	**28**	**23**	**37**	**1**
							2	**37**				

Record $(q_1, q_2, q_3, q_4, q_5, q_6, q_7, q_8)^3$
$= 9, 11, 8, 19, 29, 8, 57, 39, 47, 50, 19$
beginning from the eighth digit.

Record $p_9 = 1, 39$

Add $q_9 = 33$ to the top result-row.

2	5	39	26	22	29	28	32	52	**33**			
6	16	49	7	59	8	56	29	40				
		9	2	32	5							
	1	58	11	2	32	27	43	39				
		1	18	8	14	33	12	23	21	4	56	
			1	7	19	22	41	6	52	15	1	56
				1	27	29	2	42	1	39	50	52
					1	25	8	50	27	19	59	43
						1	37	57	28	23	37	1
							2	37	**39**	**47**	**50**	**19**
								1	**39**			

There appear to be several errors in the MS copy of the grid (see Table A.1; these have been indicated in red in our numerical recomputation):

- The leading "1" in the third row should likely be entered in the fourth row.
- The leading "1" on the tenth row should be a "2". This may explain the otherwise puzzling "2" appearing right below this.
- The terminal "49" in the tenth row should be "19".

While in essence, the procedure utilised to solve this cubic equation is broadly the same as that set out in section 3.3, there are some notable differences in both computation and layout which we outline here:

- Rows of numbers in the table are not necessarily meant to be 'read' as a single numerical value. In general, the first two significant sexagesimal digits in a row represent the current value of p_i; the remaining digits are associated with the cube of the current result-row where they differ from the previous cube. (This creates the interesting outcome of being able to 'read' the overall accurate cube of the result-row, digit by digit, along the diagonal of the table, starting with the 9 on the third row, third column.)
- Successive values p_i are generated in a manner different from that outlined in section 3.3. Rather than consider the sum using the entire current remainder, the algorithm here generates a 'refined' remainder, here denoted by r_i^*, by considering only the very next significant sexagesimal place from the original value of Crd 6°. (This produces some minor differences in the intermediate steps of working when compared with the numerical computations laid out in the reconstructed grid in section 3.3.2, but these don't affect the overall convergence.)
- In computing r_8^* at the very last step ($k = 9$), subsequent sexagesimal digits of Crd 6° have been apparently taken to be zero.

A.2 Successive-Digits Iteration for the Sine of 1°

This successive-digits iteration procedure for the sine of 1° is based on the following cubic equation:

$$x = \frac{47, 6; 8, 29, 53, 37, 3, 37, 15 + x^3}{45, 0}.$$

There is no equivalent to this exact numerical process included elsewhere in our book, namely a successive-digits iteration procedure for the sine of 1°. However, broadly speaking, as with other successive-digits procedures (albeit with different initial parameters), this iterative procedure generates a "result-row" or current approximate value $y \approx q_0, q_1, q_2, \ldots, q_k$ that is extracted digit by digit, with each digit q_i and corresponding "dividend" term p_i and "remainder" term r_i computed from the ones used in the preceding iteration.

As in the previous section, the reconstructed computational steps are presented over four columns below for each of the ten steps of the iterative procedure. Column three includes the relevant parts of the textual explanation of the computational steps and column four links these to how and where the results are recorded in the grid in Table A.2 as far as the text steps through them. Likewise, because the numerical values are aligned in a grid which preserves the absolute value of the numbers relative to each other, our reconstructions have defaulted to floating point operations for simplicity.

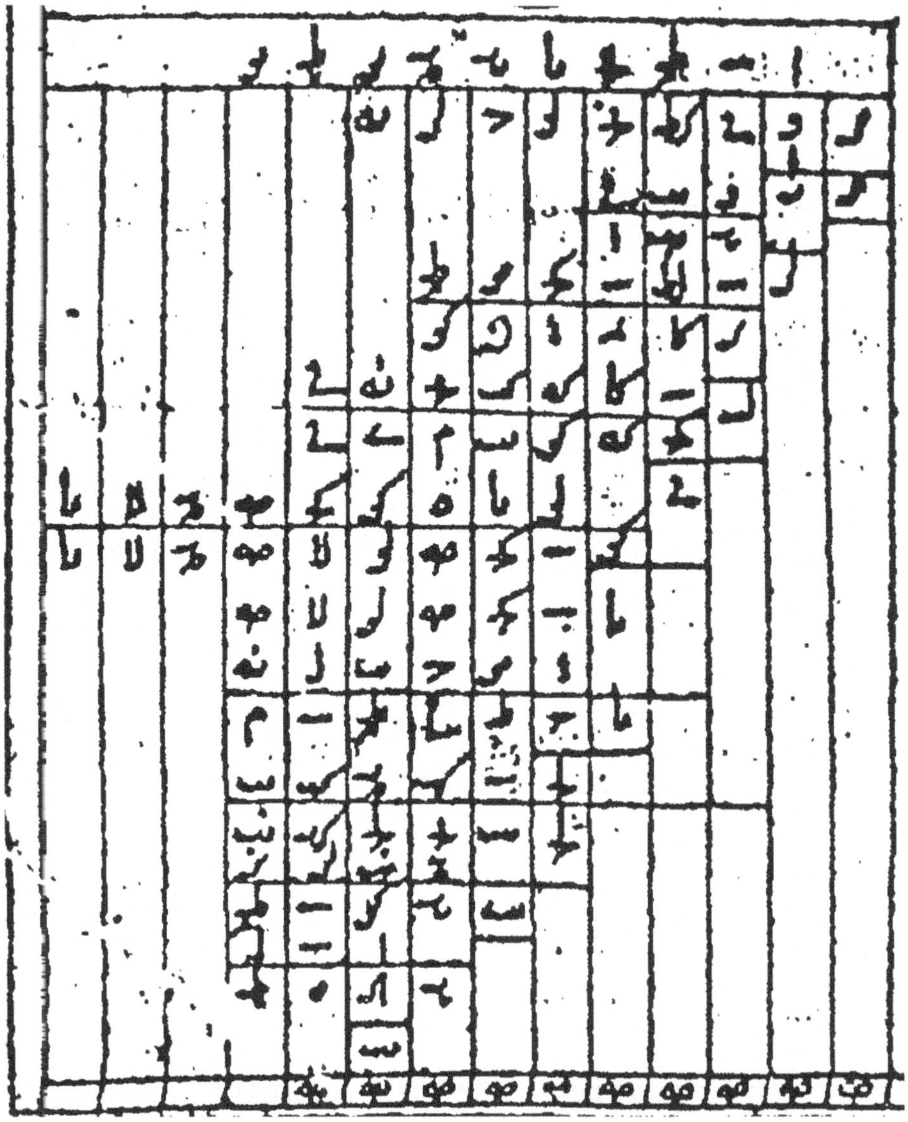

Figure A.2: Facsimile of the Tehran lithograph edition (j; p. 47) (Rosenfeld and Hogendijk 2003, p. 60) of the anonymous Arabic treatise, showing the numerical data produced from performing the prescribed iterative procedure for the sine of 1°. The value for the sine of 1° (1; 2, 49, 43, 11, 14, 16, 19, 16) appears along the top row. This grid is reproduced in table A.2.

k	p_k	q_k, r_k, and result-row $q_0, q_1, q_2, \ldots, q_k$
1	$p_0 = 15\,\text{Sin}\,3^\circ = 47, 6, 8, 29, 53, 37, 3, 37, 15$	$q_0 = p_0 \div 45$ $= 47, 6, 8, 29, 53, 37, 3, 37, 15 \div 45 \quad = 1$ $r_0 = p_0 - 45\,q_0$ $= 47, 6, 8, 29, 53, 37, 3, 37, 15 - 45 \cdot 1$ $= 2, 6, 8, 29, 53, 37, 3, 37, 15$ result-row: 1
2	$(q_0)^3 = (1)^3 \quad = 1$ r_0 : 2, 6, 8, 29, 53, 37, 3, 37, 15 + 1 $p_1 =$ 2, 7, 8, 29, 53, 37, 3, 37, 15	$q_1 = p_1 \div 45$ $= 2, 7, 8, 29, 53, 37, 3, 37, 15 \div 45 \quad = 2$ $r_1 = p_1 - 45\,q_1$ $= 2, 7, 8, 29, 53, 37, 3, 37, 15 - 1, 30$ $= 0, 37, 8, 29, 53, 37, 3, 37, 15$ result-row: 1, 2
3	$(q_0, q_1)^3 = (1, 2)^3 \quad = 1, 6, 12, 8$ 1, 6, 12, 8 − 1 0, 6, 12, 8 p_1 : +2, 7, 8, 29, 53, 37, 3, 37, 15 = 2, 7, 14, 42, 1, 37, 3, 37, 15 $45q_1$: -1, 30 p_2 : 0, 37, 14, 42, 1, 37, 3, 37, 15	$q_2 = p_2 \div 45$ $= 37, 14, 42, 1, 37, 3, 37, 15 \div 45 \quad = 0, 49$ $r_2 = p_2 - 45\,q_2$ $= 37, 14, 42, 1, 37, 3, 37, 15 - 36, 45$ $= 0, 29, 42, 1, 37, 3, 37, 15$ result-row: 1, 2, 49
4	$(q_0, q_1, q_2)^3 = (1, 2, 49)^3 \quad = 1, 8, 51, 10, 23, 46, 49$ 1, 8, 51, 10, 23, 46, 49 − 1, 6, 12, 8 0, 2, 39, 2, 23, 46, 49 p_2 : +37, 14, 42, 1, 37, 3, 37, 15 = 37, 17, 21, 4, 0, 50, 26, 15 $45q_2$: - 36, 45 p_3 : 32, 21, 4, 0, 50, 26, 15	$q_3 = p_3 \div 45$ $= 32, 21, 4, 0, 50, 26, 15 \div 45 \quad = 0, 43$ $r_3 = p_3 - 45\,q_3$ $= 32, 21, 4, 0, 50, 26, 15 - 32, 15$ $= 0, 6, 4, 0, 50, 26, 15$ result-row: 1, 2, 49, 43

"He divided 47, raised one time, by 45, also raised one time. He obtained 1 degree and the remainder from the dividend is 2 and so on." $p_0 = 47, 6, 8, 29, 53, 37, 3, 37, 15$ Place $q_0 = 1$ above in the top row of the grid (the result-row).	**1** 47 6 8 29 53 37 3 37 15
"So he wrote the cube of 1 which is also 1 degree under its kind, and he added it to the remainder, so he obtained 2 7. Then he divided the sum [by 45,0], so he obtained 2 minutes…" Record $(q_0)^3 = 1$ underneath. Record $p_1 = 2, 7$. Add $q_1 = 2$ to the top result-row.	1 **2** 47 6 8 29 53 37 3 37 15 **1** **2** **7**
"…and the remainder 37,8. Then he formed the cube of 1 2, it is 1 6 12 8, and he wrote both of them in individual [sexagesimal] places: its excess over the first cube is written under its kinds and he joined them with the remainder. It is 37 14 42 1. Then he divided this sum [by 45 0]; he obtained 49 seconds. Record difference of cubes 6, 12, 8 Record $p_2 = 37, 14, 42, 1$. Add $q_2 = 49$ to the top result-row.	1 2 **49** 47 6 8 29 53 37 3 37 15 1 2 **6 12 8** 7 **14 42 1** **37**
"The cube of 1 2 49 is 1 8 51 10 23 46 49. He took its excess over 1 6 12 8. It is 0 2 39 2 23 46 49. He wrote both (numbers) in their individual digits under their kinds and the he added them [to the remainder], so he obtained 0 17 21 4 0 50 26. Earlier from 37 degrees 15 minutes were left. He added this also, so the sum was 0 32 21 and so on. Then he divided this sum [by 45 0], and he obtained 43 thirds, and 6 seconds remained from 21 seconds, and so on." Record difference of cubes 2, 39, 2, 23, 46, 49 Record $p_3 = 32, 21, 4, 0, 50, 26$. Add $q_3 = 43$ to the top result-row.	1 2 49 **43** 47 6 8 29 53 37 3 37 15 1 2 6 12 8 7 14 42 1 **2 39 2 23 46 49** 37 17 **21 4 0 50 26** **32**

k	p_k	q_k, r_k, and result-row $q_0, q_1, q_2, \ldots, q_k$
5	$(q_0, q_1, q_2, q_3)^3 = (1, 2, 49, 43)^3$ $= 1, 8, 53, 31, 49, 9, 2, 55, 8, 7$ 1, 8, 53, 31, 49, 9, 2, 55, 8, 7 − 1, 8, 51, 10, 23, 46, 49, 0, 2, 21, 25, 22, 13, 55, 8, 7 p_3 : + 32, 21, 4, 0, 50, 26, 15, = 32, 23, 25, 26, 12, 40, 10, 8, 7 $45q_3$: - 32, 15 p_4 : 8, 25, 26, 12, 40, 10, 8, 7	$q_4 = p_4 \div 45$ $= 8, 25, 26, 12, 40, 10, 8, 7 \div 45 \quad = 0, 11$ $r_4 = p_4 - 45\,q_4$ $= 8, 25, 26, 12, 40, 10, 8, 7 - 8, 15$ $= 0, 10, 26, 12, 4, 0, 10, 8, 7$ result-row: 1, 2, 49, 43, 11
6	$(q_0, q_1, q_2, q_3, q_4)^3 = (1, 2, 49, 43, 11)^3$ $= 1, 8, 53, 32, 25, 20, 8, 21, 31, 52, 44$ 1, 8, 53, 32, 25, 20, 8, 21, 31, 52 − 1, 8, 53, 31, 49, 9, 2, 55, 8, 7 0, 0, 0, 36, 11, 5, 26, 23, 45 p_4 : + 8, 25, 26, 12, 40, 10, 8, 7 = 8, 26, 2, 23, 45, 36, 31, ‘45’ $45q_4$: - 8, 15 p_5 : 11, 2, 23, 45, 36, 31, 45	$q_5 = p_5 \div 45$ $= 11, 2, 23, 45, 36, 31, 45 \div 45 \quad = 0, 14$ $r_5 = p_5 - 45\,q_5$ $= 11, 2, 23, 45, 36, 31, 45 - 10, 30$ $= 0, 32, 23, 45, 36, 31, 45$ result-row: 1, 2, 49, 43, 11, 14
7	$(q_0, q_1, q_2, q_3, q_4, q_5)^3 = (1, 2, 49, 43, 11, 14)^3$ $= 1, 8, 53, 32, 26, 6, 11, 34, 2, 47, 56$ 1, 8, 53, 32, 26, 6, 11, 34, 2, 47 − 1, 8, 53, 32, 25, 20, 8, 21, 31, 52 0, 0, 0, 0, 46, 3, 12, 30, 55 p_5 : + 11, 2, 23, 45, 36, 31, 45 = 11, 3, 9, 48, 49, 2, 40 $45q_5$: - 10, 30 p_6 : 33, 9, 48, 49, 2, 40	$q_6 = p_6 \div 45$ $= 33, 9, 48, 49, 2, 40 \div 45 \quad = 0, 44$ $r_6 = p_6 - 45\,q_6$ $= 33, 9, 48, 49, 2, 40 - 33, 0$ $= 0, 9, 48, 49, 2, 40$ result-row: 1, 2, 49, 43, 11, 14, 44

<table>
<tr>
<td>
"...The principle was that this remainder is added to the excess of the cube of 1 2 49 43 over the cube of 1 2 49, that is to 2 21 25 22 13 55 8 7 ninths". But he added the sum of 21 and so on to this excess so he obtained 23 25 26 12 40 10 8 7. Then he subtracted 15 seconds from 23 seconds, so 8 seconds remained as is necessary. Then he divided it [by 45 0] and obtained 11 fourths, and from 25 thirds 10 thirds remained,"

Record difference of cubes: 2, 21, 25, 22, 13, 55, 8, 7

Record $p_4 = 8, 25, 26, 12, 40, 10, 8, 7$.

Add $q_4 = 11$ to the top result-row.
</td>
<td>
<table>
<tr><td></td><td>1</td><td>2</td><td>49</td><td>43</td><td>11</td><td></td><td></td><td></td><td></td></tr>
<tr><td>47</td><td>6</td><td>8</td><td>29</td><td>53</td><td>37</td><td>3</td><td>37</td><td>15</td><td></td></tr>
<tr><td></td><td>1</td><td></td><td></td><td></td><td></td><td></td><td></td><td></td><td></td></tr>
<tr><td>2</td><td></td><td></td><td></td><td></td><td></td><td></td><td></td><td></td><td></td></tr>
<tr><td></td><td></td><td>6</td><td>12</td><td>8</td><td></td><td></td><td></td><td></td><td></td></tr>
<tr><td></td><td>7</td><td></td><td></td><td></td><td></td><td></td><td></td><td></td><td></td></tr>
<tr><td></td><td></td><td>14</td><td>42</td><td>1</td><td></td><td></td><td></td><td></td><td></td></tr>
<tr><td></td><td></td><td>2</td><td>39</td><td>2</td><td>23</td><td>46</td><td>49</td><td></td><td></td></tr>
<tr><td></td><td>37</td><td></td><td></td><td></td><td></td><td></td><td></td><td></td><td></td></tr>
<tr><td></td><td></td><td>17</td><td>21</td><td>4</td><td>0</td><td>50</td><td>26</td><td></td><td></td></tr>
<tr><td></td><td></td><td></td><td>2</td><td>21</td><td>25</td><td>22</td><td>13</td><td>55</td><td>8</td></tr>
<tr><td></td><td></td><td>32</td><td></td><td></td><td></td><td></td><td></td><td></td><td></td></tr>
<tr><td></td><td></td><td></td><td>23</td><td>25</td><td>26</td><td>12</td><td>40</td><td>10</td><td>8</td></tr>
<tr><td></td><td></td><td></td><td>8</td><td></td><td></td><td></td><td></td><td></td><td></td></tr>
</table>
</td>
</tr>
<tr>
<td>
"...and then he did in their addition the same as has been explained before for seconds, so he obtained: 11 2 23 and so on, [to] the end of what he established...Treat the remaining operations in the way we have mentioned"

Record difference of cubes: 36, 11, 5, 26, 23, 45

Record $p_5 = 11, 2, 23, 45, 36, 31, 45$.

Add $q_5 = 14$ to the top result-row.
</td>
<td></td>
</tr>
</table>

k	p_k	q_k, r_k, and result-row $q_0, q_1, q_2, \ldots, q_k$
8	$(q_0, q_1, q_2, q_3, q_4, q_5, q_6)^3$ $= (1, 2, 49, 43, 11, 14, 44)^3$ $= 1, 8, 53, 32, 26, 8, 36, 18, 24, 59, 57$ 1, 8, 53, 32, 26, 8, 36, 18, 24, 59 − 1, 8, 53, 32, 26, 6, 11, 34, 2, 47 0, 0, 0, 0, 2, 24, 44, 22, 12 p_6 : + 33, 9, 48, 49, 2, 40 = 33, 12, 13, 33, 24, 52 $45q_6$: - 33 p_7 : 12, 13, 33, 24, 52	$q_7 = p_7 \div 45$ $= 12, 13, 33, 24, 52 \div 45 \quad = 0, 16$ $r_7 = p_7 - 45\,q_7$ $= 12, 13, 33, 24, 52 - 12, 0$ $= 0, 13, 33, 24, 52$ result-row: 1, 2, 49, 43, 11, 14, 44, 16
9	$(q_0, q_1, q_2, q_3, q_4, q_5, q_6, q_7)^3$ $= (1, 2, 49, 43, 11, 14, 44, 16)^3$ $= 1, 8, 53, 32, 26, 8, 37, 11, 2, 57, 7$ 1, 8, 53, 32, 26, 8, 37, 11, 2, 57 − 1, 8, 53, 32, 26, 8, 36, 18, 24, 59 0, 0, 0, 0, 0, 0, 52, 37, '57' p_7 : + 12, 13, 33, 24, 52 = 12, 14, 26, 2, 49 $45q_7$: - 12 p_8 : 14, 26, 2, 49	$q_8 = p_8 \div 45$ $= 14, 26, 2, 49 \div 45 \quad = 0, 19$ $r_8 = p_8 - 45\,q_8$ $= 14, 26, 2, 49 - 14, 15$ $= 0, 11, 2, 49$ result-row: 1, 2, 49, 43, 11, 14, 44, 16, 19
10	$(q_0, q_1, q_2, q_3, q_4, q_5, q_6, q_7, q_8)^3$ $= (1, 2, 49, 43, 11, 14, 44, 16, 19)^3$ $= 1, 8, 53, 32, 26, 8, 37, 12, 5, 27, 11$ 1, 8, 53, 32, 26, 8, 37, 12, 6, 27 − 1, 8, 53, 32, 26, 8, 37, 11, 2, 57 0, 0, 0, 0, 0, 0, 1, 2, 30 p_8 : + 14, 26, 2, 49 = 14, 27, 5, 19 $45q_8$: - 14, 15 p_9 : 12, 5, 19	$q_9 = p_9 \div 45$ $= 12, 5, 19 \div 45 \quad = 0, 16$ result-row: 1, 2, 49, 43, 11, 14, 44, 16, 19, 16

	1	2	49	43	11	14	44	16	19	16			
47	6	8	29	53	37	3	37	15					
	1												
2													
		6	12	8									
	7												
		14	12	1									
		2	39	2	23	46	49						
	37												
		16	21	4	0	50	26						
			2	21	25	22	13	55	8				
		32											
			23	25	26	12	40	10	8				
					36	11	5	26	23	45	44	31	11
			8										
				26	2	23	45	36	31	45	44	31	11
					2	23	45	36	31	45			
				11									
					0	46	3	12	30	55			
				11	3	9	48	49	2	40			
					33	2	24	44	22	12			
					33	12	13	33	24	52			
							0	52	37	57			
						12	14	26	2	49			
								1	2	30			
							14	27	5	19			
								12					
45	45	45	45	45	45	45	45	45	45				

Table A.2: Translation of the numerical grid in figure A.2.

Three minor errors appear in the MS copy of al-Kāshī's grid:

- the 12 in the value 37,14,12,1,37,3,37,15 should read 42;
- the 16 in the value 37,16,21,4,0,50,26,15 should read 17;
- a row later in the table is accidentally duplicated.

The two incorrect values are both easily explained as scribal errors when using *abjad* numeration.

Some notable features of this procedure and layout are as follows:

- Rows of numbers in the grid are not necessarily meant to be read as a single numerical value.
- There are sometimes level of precision differences between the numerical values recorded in the text and in the tables.
- There is a minor difference in order of operations in this successive-digits approach. Here the process proceeds by adding the difference of cubes before subtracting the amount to be taken from the remainder.

Given the complexity of the iterative process and its layout in the grid included in the manuscript, for the reader's comprehension, we include here the sequence of reconstructed computations recorded in a grid (see table A.3) similar to that appearing in this anonymous Arabic text. In our grid of reconstructed values, the differences of cubes to be added to the remainder are displayed in red, and the remainder of the quotient to be subtracted from the remainder (which do not appear in al-Kāshī's grid) are displayed in blue.

	1	2	49	43	11	14	44	16	19	16			
47	6	8	29	53	37	3	37	15					
	1												
47	7	8	29	53	37	3	37	15					
2	0												
2	7	8	29	53	37	3	37	15					
		6	12	8									
2	7	14	42	1	37	3	37	15					
1	30												
	37	14	42	1	37	3	37	15					
		2	39	2	23	46	49						
	37	17	21	4	0	50	26	15					
	36	45											
		32	21	4	0	50	26	15					
			2	21	25	22	13	55	8				
		32	23	25	26	12	40	10	8				
		32	15										
			8	25	26	12	40	10	8				
					36	11	5	26	23	45	44	31	11
			8	26	2	23	45	36	31	45	44	31	11
			8	15									
				11	2	23	45	36	31	45			
						46	3	12	30	55			
				11	3	9	48	49	2	40			
				10	30								
					33	9	48	49	2	40			
						2	24	44	22	12			
					33	12	13	33	24	52			
					33								
						12	13	33	24	52			
								52	37	57			
						12	14	26	2	49			
						12							
							14	26	2	49			
								1	2	30			
							14	27	5	19			
							14	15					
								12	5	19			
45	45	45	45	45	45	45	45	45	45				

Table A.3: Reconstruction of the numerical values and their intermediate amounts recorded in figure A.2.

Bibliography

Aaboe, Asger (1954). "Al-Kashi's iteration method for the determination of $\sin 1°$." In: *Scripta Mathematica* 20, pp. 24–29.

Aydin, Nuh and Lakhdar Hammoudi (2015). "Root extraction by Al-Kashi and Stevin." In: *Archive for History of Exact Sciences* 69, pp. 291–310.

– (2019). *Al-Kāshī's Miftāḥ al-Ḥisab, Volume I: Arithmetic*. Cham: Birkhäuser.

Azarian, Mohammad K. (2015). "A study of Risāla al-Watar wa'l Jaib ("The Treatise on the Chord and Sine)." In: *Forum Geometricorum* 15, pp. 229–242.

Berggren, J.L. (1986). *Episodes in the Mathematics of Medieval Islam*. New York: Springer.

Chaturveda, Muralidhar, ed. (1976). *Samrāṭsiddhānta of Jagannātha Samrāṭ*. Sagar: Samskrita Parishat, Saugar University.

Dvivedi, Sudhakara and Muralidhar Jha, eds. (1991). *Siddhāntatattvaviveka of Kamalākara*. Varanasi: Chaukhamba Surbharati Prakashan.

Gerl, Armin (1989). *Trigonometrisch-Astronomisches Rechnen Kurz Vor Copernicus: Der Briefwechsel Regiomontanus-Bianchini*. Stuttgart: Franz Steiner Verlag.

Gold, David and David Pingree (1991). "A hitherto unknown Sanskrit work concerning Mādhava's derivation of the power series for sine and cosine." In: *Historia Scientiarum* 42, pp. 49–65.

Gupta, R. C. (1969). "Second order interpolation in Indian mathematics up to the fifteenth century." In: *Indian Journal of History of Science* 4, pp. 86–98.

Hayashi, Takao (1991). "A note on Bhāskara I's rational approximation to Sine." In: *Historia Scientiarum* 42, pp. 45–48.

– (1997). "Āryabhaṭa's rule and table for Sine-differences." In: *Historia Mathematica* 24, pp. 396–406.

Islamic Scientific Manuscripts Initiative (n.d.). URL: `https://ismi.mpiwg-berlin.mpg.de/`.

Kasir, D. S. (1972). *The Algebra of Omar Khayyam*. New York: AMS Press.

Kusuba, Takanori and David Pingree (2002). *Arabic Astronomy in Sanskrit: Al-Birjandī on Tadhkira II, Chapter 11 and its Sanskrit Translation*. Leiden: Brill.

Mercier, Raymond (1993). "Account by Joseph Dubois of astronomical work under Jai Singh Sawā'ī." In: *Indian Journal of History of Science* 28, pp. 157–66.

Misra, Anuj (2016). "The *Golādhyāya* of Nityānanda's *Sarvasiddhāntarāja*: An Examination of 'The Chapter on Spheres' in a Seventeenth Century text on Mathematical Astronomy." PhD thesis. University of Canterbury.

Montelle, Clemency (2019). "The Production of Sines: Bhāskarācārya's *Jyotpatti*." In: *Bhāskara-prabhā*. Ed. by Takao Hayashi, Clemency Montelle, and K. Ramasubramanian. Culture and History of Mathematics 11. Hindustan Book Agency, pp. 291–300.

Montelle, Clemency and K. Ramasubramanian (2018). "Determining the Sine of one degree in the *Sarvasiddhāntarāja* of Nityānanda." In: *SCIAMVS* 19, pp. 1–52.

Pingree, David (1999). "An astronomer's progress." In: *Proceedings of the American Philosophical Society* 143, pp. 73–85.

– (2003). *A Descriptive Catalogue of the Sanskrit Astronomical Manuscripts preserved at the Maharaja Man Singh II Museum in Jaipur, India*. Philadelphia: American Philosophical Society.

– ed. (2004). *Siddhāntakaustubha of Jagannātha Paṇḍita*. New Delhi: Indian National Science Academy.

Plofker, Kim (1996). "An example of the secant method of iterative approximation in a fifteenth-century Sanskrit text." In: *Historia Mathematica* 23, pp. 246–256.

– (2004). "The problem of the sun's corner altitude and convergence of fixed-point iterations in medieval Indian astronomy." In: *Studies in the History of the Exact Sciences in Honour of David Pingree*. Ed. by Charles Burnett et al. Leiden: Brill, pp. 552–586.

– (2009). *Mathematics in India*. Princeton: Princeton University Press.

C. Montelle et al., *From Samarqand to Jaipur*, Sources and Studies in the History of Mathematics and Physical Sciences, https://doi.org/10.1007/978-3-031-89087-1

Ramasubramanian, K. and M. S. Sriram (2011). *Tantrasaṅgraha of Nīlakaṇṭha Somayājī*. Culture and History of Mathematics 6. New Delhi: Hindustan Book Agency.

Rosenfeld, B. A. and J. P. Hogendijk (2003). "A mathematical treatise written in the Samarqand observatory of Ulugh Beg." In: *Zeitschrift für Geschichte der arabisch-islamischen Wissenschaften* 15, pp. 25–65.

Sarma, K.V. et al. (2008). *Gaṇita-Yukti-Bhāṣā (Rationales in Mathematical Astronomy) of Jyeṣṭhadeva*. London: Springer.

Śāstrī, Bāpū Deva (1989). *The Siddhāntaśiromaṇi, a Treatise on Astronomy*. 2nd edition. Kashi Sanskrit Series 72. Varanasi: Chaukhambha Sanskrit Sansthan.

Sédillot, Louis Amélie (1839). *Tables astronomiques d'Oloug Beg, commentées et publiées avec le texte en regard*. Tome I, 1 fascicule. Paris.

– (1853). *Prolégomènes des tables astronomiques d'Oloug-Beg: Traduction et Commentaire*. Paris: F. Didot.

Sharma, Ram Charan (1986–1994). *Vāṅmaya-cayanikā: A Classified Catalogue of Selected Manuscripts from Universal Institute of Orientology and Museum of Indology*. 6 vols. Jaipur: Universal Institute of Orientology and Museum of Indology.

Sharma, Ram Swarup, ed. (1967–1969). *Samrāṭsiddhānta (Siddhāntasārakaustubha) of Jagannātha*. New Delhi: Indian Institute of Astronomical and Sanskrit Research.

Sharma, Virendra Nath (1995). *Sawai Jai Singh and His Astronomy*. Delhi: Motilal Banarsidass.

Toomer, G.J. (1984). *Ptolemy's Almagest*. London: Gerald Duckwork & Co. Ltd.

Van Brummelen, Glen (2009). *The Mathematics of the Heavens and the Earth: The Early History of Trigonometry*. Princeton: Princeton University Press.

– (2021). *The Doctrine of Triangles: A History of Modern Trigonometry*. Princeton, NJ: Princeton University Press.

– (n.d.). "Crossing a mathematical Rubicon: Al-Kashi's two methods of computing Sin(1 °)." In: *Mathematical Practices in Astral Sciences*. Ed. by Matthieu Husson and Karine Chemla.

Van Brummelen, Glen, Taro Mimura, and Yousuf Kerai (2013). "Al-Samaw'al's curious approach to trigonometry." In: *Suhayl* 11, pp. 9–31.

van Gent, Robert Harry (n.d.). *The Astronomical Tables of Ulugh Beg*. URL: https://webspace.science.uu.nl/~gent0113/ulughbeg/ulughbeg_zij_i_sultani.htm.

Voss, Don L. (1985). "Ibn al-Haytham's 'Doubts Concerning Ptolemy: A Translation and Commentary'." PhD thesis. University of Chicago.

The manufacturer's authorised representative in the EU is Springer Nature Customer Service Centre GmbH, Europaplatz 3, 69115 Heidelberg, Germany. If you have any concerns regarding our products, please contact ProductSafety@springernature.com

Printed and bound by CPI Group (UK) Ltd, Croydon, CR0 4YY
30/09/2025
02225516-0015